This book delivers a quantitative account of the science of cosmology, designed for a non-specialist audience. The basic principles are outlined using simple maths and physics, but still providing rigorous models of the universe. It offers an ideal introduction to the key ideas in cosmology, without going into technical details. The approach used is based on the fundamental ideas of general relativity, such as the spacetime interval, comoving coordinates, and spacetime curvature. It provides an up to date and thoughtful discussion of the big bang and the crucial issues of structure and galaxy formation. Questions of method and philosophical approaches in cosmology are also briefly discussed. Advanced undergraduates in either physics or mathematics would benefit greatly from the book's use either as a course text or as a supplementary guide to cosmology courses.

Cosmology: a first course

Cosmology

A First Course

MARC LACHIÈZE-REY

TRANSLATED BY JOHN SIMMONS

CAMBRIDGE
UNIVERSITY PRESS

Published by the Press Syndicate of the University of Cambridge
The Pitt Building, Trumpington Street, Cambridge CB2 1RP
40 West 20th Street, New York, NY 10011-4211, USA
10 Stamford Road, Oakleigh, Melbourne 3166, Australia

Originally published in French as *Initiation à la cosmologie* by Masson, Paris and © Masson 1992

First published in English by Cambridge University Press 1995 as *Cosmology: a first course*

© Cambridge University Press 1995

English translation printed in Great Britain at the University Press, Cambridge

A catalogue record of this book is available from the British Library

Library of Congress cataloguing in publication data

Lachièze-Rey, Marc.
Connaissance du cosmos. English
Cosmology : a first course / Marc Lachièze-Rey, translated by John Simmons.
   p. cm.
Includes bibliographical references and index
ISBN 0-521-47966-5
1. Cosmology.   2. General relativity (Physics)
I. Title.
QB981.L2313   1995
523.1–dc20   94-46437   CIP

ISBN 0 521 47441 8 (hardback)
ISBN 0 521 47966 5 (paperback)

TAG

Contents

1

Discovering the cosmos

Can one study the universe?

The study of the universe or cosmos, cosmology, is the most global and all embracing study possible. But does the universe actually exist as a meaningful entity? The notion of the cosmos involves more than the collection of the diverse heavenly bodies under the umbrella of a single name. It implies the existence of global and universal properties and relations. For this reason the cosmologist is not interested in the different astronomical objects populating the universe for their own sake, whether they are planets, stars, galaxies or whatever. The cosmologist is interested in their mutual relationships, the framework within which they evolve, the general laws governing them and the overall structure that contains them. It is this totality that constitutes the global properties of the cosmos.

Astrophysical phenomena are of interest to the cosmologist only in so far as they have a cosmological significance. It is this that provides the link with astrophysics and astronomy.

Take for example the distribution of galaxies. This interests the cosmologist in several ways. In the first place the dynamical processes producing this distribution, and responsible for the formation of galaxies and structure, are part of cosmic history. But galaxies also play the role of markers or pointers that make it possible for us to unravel the geometric structure of the universe itself. This concept, which we shall develop later, has to be seen in the light of the general theory of relativity. To analyse structure we must study galaxies, even if only to measure their distance. Thus the cosmologist cannot avoid being both astronomer and astrophysicist.

The term cosmology now embraces what was previously divided into two distinct fields, cosmography and cosmogony.

1

Cosmography, which is certainly the branch furthest removed from astrophysics, is the study of the global geometric structure of spacetime (with or without curvature, finite or infinite, homogeneous or not, isotropic or not), without any reference to the material content of spacetime. It makes use of both special and general relativity, and draws upon the mathematics pertaining to these theories.

All the observations are in agreement with an evolving cosmos, and of course the expansion of the universe is one very obvious manifestation of this evolution. It is this structure and evolution of the universe that cosmological models attempt to describe. Physical processes take place within this cosmological setting – the progressive appearance of physical objects such as atoms, molecules, stars and galaxies and their organisation. Some of these phenomena are similar to those physicists have been able to observe in the laboratory and under specific conditions – expansion and cooling, the density dilution of matter etc. Cosmogony studies this process of creation. Cosmogony is much more rooted in physical processes. It includes the study of the matter in the universe, its evolution throughout cosmic history, and addresses the question of how to observe these cosmological phenomena. This physical cosmology draws from every field of physics, including particle physics. Obviously these two branches of cosmology are intimately linked.

Is a scientific cosmology possible?

This question arises because of the uniqueness of the universe as a cosmological object. There is only one universe, and we are part of it. This situation is different from the one usually encountered in science. Hence the methods and concerns of cosmology are distinct from the rest of physics.

In carrying out 'normal' science it is necessary to be able to reproduce and repeat scientific experiments and investigations under specific and well defined conditions. This is the essence of the experimental method. Cosmology is different. When cosmologists construct a particular cosmological model, it is impossible to create a corresponding universe in order to see how it might evolve! Is there even any sense in talking about the possible existence of such a universe? Cosmology has to infer its laws from just one sample.

This discipline can nevertheless claim to be a science, and moreover a natural science based on observation. Cosmological models, including the big bang, are founded, at least partially, on observations. But above all cosmology starts out from a number of fundamental principles. In this sense it is an abstract science: the history of the universe is a construction of the mind. If confirmed by observation, it is only in a relatively indirect fashion.

Cosmology is based on definite principles and can be compared with observations. It is a science. However it is a science that is all the more special because of the nature of the object it sets out to study, and because of the special position of the observer in relation to it. The astronomer, the cosmologist, and each one of us, exists within the object that he or she studies. As a consequence prejudices arise, or as astronomers say, 'biases', which affect our interpretation of the observations.

As with all the sciences, though no doubt to a larger degree, cosmology rests on myths. Perhaps it is still an imperfect science in so far as the prejudices affecting it and the influences of the various different myths are difficult to recognise and uncover.

In studying the fundamental problems concerning the nature of our universe, its origin and its future, as well as the place of humanity in this universe, cosmology shares its principal interests with disciplines that are outside the province of science: metaphysics, theology and philosophy. Thus it is not always easy to draw a line between science and metaphysics. Nor is it easy to recognise the influences of the different myths to which we are subjected or the influences of our cultural and social heritage. In any case it is important to be aware that scientific cosmology does rest on several metaphysical principles.

Principles of cosmology

Without doubt one of the first principles of cosmology is the statement that a scientific cosmology has a meaning and that a universe that does not reduce to a simple accumulation of existing objects is, as such, understandable. Indeed it is this, as we have said, that constitutes cosmology's 'raison d'être' and defines the object it sets out to study.

On the other hand, to compare the various cosmological models with observations it is necessary to observe and establish the nature of the most distant astronomical objects. Thus it is necessary to have a knowledge of the phenomena that take place in these regions of the universe and of the physics that holds there. Astronomers and cosmologists assume that the laws of physics are the same everywhere; the laws of gravitation and electromagnetism, of quantum physics etc. are the same at every point of the universe. How could one think otherwise? Without this article of faith no scientific cosmology is possible. But once again it is a question of accepting an a priori principle. In fact, the very idea of a physical law implies its universality. But a law is only valid within a certain domain (for example of temperature or of energy). Often, however, the physical conditions in the universe lie outside this domain.

Most cosmologists go still further. They profess a belief that the universe is homogeneous, that is it appears the same at every point of space. This cosmological principle is also a dogma, and forms the basis for the most popular models of the universe. Yet the few available observations are so fragmentary compared with the scale of the cosmos that they cannot possibly establish the validity of such a principle. It states that there is no privileged point in the cosmos! The big bang models, which are today considered as 'standard', assume this homogeneity from the outset. An examination of the recent literature reveals a curious situation: certain authors appear to be surprised by the homogeneity of the universe, and attempt to explain this fact within the framework of the same big bang models. However, in such models it is postulated from the beginning as a principle! This contradiction shows that the essence of these models is not always understood.

If the cosmos is indeed the same at every point in space, one might ask whether it has always been the same throughout its history. This principle, which has seduced several cosmologists, goes under the name of the perfect cosmological principle. It is no less reasonable than the previous principle. However, observations, first and foremost universal expansion, appear to show that the universe evolves. It is practically impossible to reconcile the idea of a stationary universe with observations.

The scope of cosmology

The advancement of technique broadens our vision of the universe. The boundaries of the universe recede yet still remain inaccessible. The cosmos remains a mysterious. Mankind has always attempted to associate ideas, concepts and indeed deities within this framework. Etymologically speaking the cosmos was associated with order and harmony, and was long thought to be the refuge of the gods. Today the cosmos is seen as a field of application for the advancement of science. Yet can we also detect signs of a discipline in the process of deification? In any case, it is certain that the cosmos remains an object of fascination whether one considers it to be rational, material, and neutral, or whether one sees in the history of the universe some cosmic intention at work.

My intention here is not to discuss these exciting questions, which I leave to the philosophers and epistemologists (see, for example, the works of J. Merleau-Ponty), but to present the 'models' that physicists and astrophysicists consider adequate explanations of the cosmos and its evolution. I shall not try to unravel the part played by mythology. I shall leave it up to the reader to recognise the influence of this or that mythological concept.

If we accept that the universe possesses and indeed defines itself through a structure, we are in a position to be able to state the essential aim of cosmology: to determine both this structure and its history and evolution. Is the universe closed or infinite? Does it have a beginning? Will there be an end?

In order to answer these questions, we must examine the processes that take place in the most distant celestial bodies and understand the physics governing these processes so that we may interpret the observations. This requires the application of the laws of physics to these objects. We have to postulate that these are identical to those we already know: gravitation, electromagnetism, quantum physics etc., and are the same everywhere in the universe (and ultimately hold for all time, although we shall discuss this in more depth later).

The properties of the cosmos influence those of the matter it contains. Galaxies, set in the geometry of the universe, are subject to its evolution. Indeed they reveal the expansion, the structure and evolution of the universe. Some of their properties, such as expansion, have a cosmic character. But the properties of the cosmos are also influenced (if not totally determined) by its material contents. For this reason, there is no possible cosmology without the study of the properties of galaxies, or without astronomy and astrophysics. It is this that gives cosmology the status of an observational science.

1.1 The universe as it appears

Knowing, at least partially, the material contents of the universe is one of the first steps in studying it. In this respect cosmology is an off-shoot of astrophysics.

1.1.1 From the Earth to the stars

The Earth is just one of nine planets of the Solar System. This tiny island occupies a minute volume of the universe. The typical terrestrial distances of a few thousand kilometres are only a fraction of a parsec*, which is the unit of distance used by astronomers (one parsec is equal to 3×10^{13} km or 3.262 light-years). If we are to study the universe, on the other hand, we shall have to deal with distances of millions of parsecs or megaparsecs (Mpc). The Sun is a fairly ordinary star among thousands of others, and microscopic on a cosmological scale.

Beyond the Solar System a few tens of thousands of millions of stars are clustered together to form the Milky Way*, our Galaxy. Each star is similar to the Sun, but can have a different mass, age, luminosity, colour or chemical composition etc. (Astronomers still do not know whether all stars have planetary systems like the Sun or not.) Typically there is a separation of several parsecs between stars, which means that effectively each planetary system can be considered as a point. The ensemble of stars, our Galaxy, forms a system with a diameter of a few tens of thousands of parsecs. Beyond this there is intergalactic space.

It is only much further out, at distances of megaparsecs, that one comes across matter again in other galaxies. The universe is filled with galaxies more or less like our own, and each one an 'island universe'.

On the scale of the universe, these galaxies, of which astronomers have observed several tens of thousands of examples, can be considered to be point-like! The cosmos is filled with a 'gas of galaxies'. These galaxies are separated from each other by an average distance of a megaparsec. For the cosmologist they constitute the fundamental objects of the universe.

Astrophysicists attempt to trace the evolution of these galaxies, to understand how they are formed and how and why they are distributed as they are. Further out, and amongst these galaxies, there are objects remarkable for the enormous quantity of energy that they emit. These are the quasars, which belong to a class of objects called active galactic nuclei (or AGNs). Their nature is still a mystery, and we shall discuss them later. All these objects we shall designate generically by the term galaxy.

Cosmologists are interested in all these objects, not only in their individual properties, but in their evolution and how they are distributed, or in short, in their collective properties and their relation to the 'cosmos'. Effectively they give us a window onto the universe and its structure. If one cannot 'see' directly the geometry or evolution of the cosmos, one can reconstruct it by observing galaxies, which play the role of tracers or markers. It is up to the cosmologist to recognise and then isolate those properties with a universal character. This is the principle underlying 'cosmological tests'. The best definition of the cosmos is possibly the ensemble of relations between the constituent parts of the cosmos.

1.1.2   Stars and luminous matter

Stars are the basic units of luminous matter. They are made of 'baryonic matter'*: protons and neutrons (to which are associated the electrons). Stars are held together primarily by gravitation which gives them their spherical form. It is the internal pressure of the star that counteracts the gravitational attraction and so prevents its total collapse.

The internal pressure consists of both the thermal pressure of the gas making up the star and radiation pressure resulting from the nuclear reactions in the stellar core. This sort of equilibrium, between a gravitational force and an opposing interaction, is a general phenomenon in the cosmos.

One finds it at every level, from planets to clusters of galaxies, and even on the scale of the universe itself.

Stars are active. Nuclear reactions in their cores transform one chemical species to another, and are responsible for the luminosity we observe. In the core of the Sun, for example, protons are transformed into helium* nuclei (a process that goes under the name of hydrogen burning).

During the major part of its life a star undergoes hydrogen burning in its core. It is then said to be on the main sequence. At the end of a few thousand million years (10^{10} years for the Sun), another cycle of reactions ensues and the nature of the star changes.

The details of star formation are still poorly understood. A cloud of cold gas condenses under its own weight. During this contraction it heats up and emits electromagnetic radiation. This energy carries away part of the energy of the cloud, and thus accelerates condensation etc. In all likelihood, large clouds fragment into smaller clouds, each condensing and heating up.

After a certain time, this small cloud takes a spherical form. It has become so hot that there is an onset of nuclear reactions in its core. These reactions release energy which is transformed into light and emitted; a star is born. Meanwhile gas and radiation pressure have increased sufficiently to counterbalance gravitational attraction. The star ceases to contract and remains in equilibrium, like the Sun. This equilibrium will last until nuclear reactions stop, when the nuclear combustible material is used up, or are modified.

1.1.3 Our Galaxy and other galaxies

Most stars we observe, the nearest and thus the brightest, belong to our Galaxy. The mean density of matter in our Galaxy in the form of stars is $2 \times 10^{-24}\,\mathrm{g\,cm^{-3}}$ compared with a mean density for the universe of about $10^{-31}\,\mathrm{g\,cm^{-3}}$: the density contrast

$$\delta = \frac{\delta\rho}{\rho}$$

being around 10^7.

Not all the stars of our Galaxy are the same. Some are old, low mass and low luminosity stars. These are mainly cool and red, and distributed in our Galaxy in the form of an almost spherical halo; they constitute what is called population II stars. Population I stars on the other hand are young blue stars and are found in the Galactic disc, forming the spiral arms of the Galaxy.

All the visible matter of the universe is found in galaxies, our own being just one of thousands of millions. The luminosity of each of these galaxies

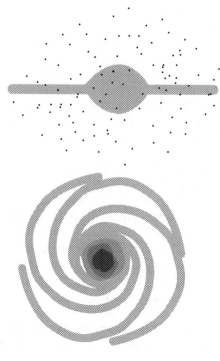

(a)

(b)

Figure 1.1: The structure of our Galaxy as it would appear (a) in profile and (b) from above.

is the total luminosity of all the stars it contains. Galaxies were classified by Hubble (1936) into different types or classes according to essentially morphological criteria.

In the spirals, like our own, there are thousands of millions of stars in the form of a flattened disc within which lie the spiral arms (and often other structures such as bars and rings). These galaxies also contain a lot of gas from which new stars form. They rotate about the axis of the disc. In order to explain the rate of this rotation astronomers invoke the presence of a great quantity of invisible matter, in addition to the matter contained within the stars. This is the so called problem of dark matter*.

Elliptical galaxies are essentially different by virtue of their form. The stars contained in them are much older, and scatter within an ellipsoid. They contain much less gas, and their dynamical behaviour is also different. Lenticular, or SO, galaxies have properties that are intermediate between spiral and elliptical.

In addition to the Hubble types, one must add the dwarf and giant irregular galaxies, as well as the various classes of active galaxies that emit huge quantities of energy through means that do not involve stars.

The link between these different classes is still poorly understood, and poses an enigma that will have to be resolved by models of galaxy formation. Do spirals and ellipticals really form in different conditions, or do the latter evolve from the former? At least in certain cases it is clear that the collision of two spirals can result in the formation of a giant spiral. Is this also the general case?

Galaxy formation is still poorly understood. Astronomers do not know what a newly formed galaxy or one that is in the process of forming looks like.

1.1.4   Primordial galaxies and protogalaxies

One sign that a galaxy is young is the presence of very young, in the sense of recently formed, bright, blue (owing to their high surface temperature) stars. Another, observable essentially in the infrared, is the presence of gas and dust clouds that are in the process of collapsing.

It is certain that galaxies have not always been around in the universe. The first did not appear before one thousand million years of cosmic history had elapsed. When did they form, and what were the first galaxies like? We still do not know. Probably they formed from clouds contracting under their own weight to form protogalaxies.

Protogalaxies, by definition, contain no stars. They are nevertheless observable. Since they do not and never did contain stars, all those chemical elements uniquely produced inside stars must be absent. Hydrogen and helium are virtually the only elements present. So protogalactic clouds should in the future be identified as protogalaxies.

Protogalaxies become actual galaxies once the stars have begun to form and radiate. Although the form of such partially formed or primordial galaxies was undoubtedly very different from any today, we do not know what they looked like. However, astronomers are actively looking for them. Obviously it is necessary to look far back in time, and far out into space. These objects will be observable, but only with difficulty.

1.2   The distribution of matter

We have known since the beginning of the century that matter exists in the form of galaxies. Indeed this has been suspected since the eighteenth century. Galaxies are the main objects studied in cosmology. They are the largest objects whose nature we understand. Most importantly we believe that their properties are explained more by cosmic processes than by local conditions. Unlike planets and stars which live in a local environment and have no direct relationship with the rest of the universe, galaxies must be seen in the cosmological setting.

Andromeda, which is one of our nearest neighbours, was one of the first galaxies to be recognised. In our neighbourhood there is a small group of galaxies, going under the name of the Local Group (LG)*, which contains Andromeda and about fifteen smaller galaxies in addition to the Milky Way, our Galaxy. They mutually interact gravitationally and are in a volume of $1\,Mpc^3$. The majority of galaxies observed today are situated well beyond the LG.

1.2.1   Beyond the Local Group

In order to understand its structure, astronomers map out the universe. In doing this they forget the individual properties of the galaxies and concentrate only on their distribution in space. The position on the celestial sphere, defined for example by the angular coordinates right ascension* and declination*, provide the first bits of information. But this is only the 'projection' onto the celestial sphere. To study the actual spatial distribution one needs to know distances, and this is extremely difficult in astronomy, as we shall see in sections 1.3 and 2.2.6.

There are innumerable galaxies, and the task, undertaken since the beginning of the century, of drawing up catalogues of galaxies is enormous. Thanks to the continual improvement in telescopes, detectors and instrumentation, important progress has been achieved in recent times.

1.2.2   Clusters and superclusters

Maps clearly reveal that galaxies are not distributed randomly but are clustered in groups and clusters. The Local Group is the closest example. Beyond, at about 15 Mpc, is the Virgo Cluster, which is much richer than the Local Group, yet smaller than the average cluster. Beyond Virgo there are many more clusters. They were first catalogued in 1958 by George Abell from photographic plates obtained from the Mount Palomar Telescope.

Within the nearest tens of megaparsecs matter displays a still larger structure, the so called Local Supercluster*, which contains eleven clusters and several field galaxies. It is flat and about one megaparsec thick. It is also called the Virgo supercluster.

The distribution of galaxies and clusters is not uniform. Superclusters are the rule at large scales. They are typically long and flat, like our own, have length scales characteristically of twenty or so megaparsecs, and are possibly linked together by huge bridges or filaments of matter.

If the superclusters appear as condensations of matter at very large scales, there also exist immense voids that are free of matter. One such void has been detected in the constellation of Boötes. It is at a distance of around 150 Mpc from us, and about 30 Mpc across. Voids on these and smaller scales appear quite common.

Thus the distribution of galaxies appears very irregular. The large observed structures, flattened or elongated superclusters, seem to correspond to the bounding surfaces of giant voids. So we have a vast network, an immense cellular structure in two rather than three dimensions. The latest manifestation of this structure is a Great Wall due to the presence of interlacing sheets.

This structure, still only partially pieced together, is being progressively revealed as observations are refined and extended. Astronomers do not know how exactly to describe it. Is it better to talk of a 'sponge-like structure' or 'bubbles', of a 'net' or 'cells' and 'sheets'? Do clusters correspond to the intersection of filaments or sheets? Our inability to describe this structure masks a much more serious failure: we still have no idea of the processes that have led to this structure formation. Great advances are being made in this field, both from the point of view of observations and of theory (see chapter 5 on the formation of galaxies).

Is the maximum length scale of around 25 Mpc of the known structures a product of our limited capacity in mapping out larger structures, or is it a real limit on structure formation in the cosmos?

1.2.3 Catalogues

Most results have been obtained through a three dimensional analysis of the distribution of galaxies. To carry out such an analysis requires a knowledge of the distances of galaxies, which can be estimated, as we show in the next chapter, from redshift measurements, although this process is relatively lengthy. A large part of the results come from the analysis of the CfA catalogue of the Center for Astrophysics at Cambridge, Massachusetts, USA, where astronomers have undertaken a very long observational programme to estimate the distances of the greatest possible number of galaxies. Other catalogues have been drawn up, notably in the southern hemisphere, as well as other deeper surveys aimed at uncovering the distribution of galaxies at the most distant scales. It was due to one such survey that the giant void in Boötes was discovered.

The processes that have led to the formation of galaxies must by the same token be responsible for their spatial distribution and it is they that the catalogues attempt to identify. In order to do this it is first necessary to express quantitatively the properties of the distribution of galaxies, and this requires specific statistical tools adapted for the job.

In relation to the formation of galaxies, we shall see that one can distinguish two regimes in their distribution. At the largest scales (above 8 Mpc approximately), dynamic processes certainly have not been able to act effectively. They have not gone beyond the so called linear stage, and one refers to the distribution on large scales as the linear domain. In contrast, on smaller scales, roughly speaking smaller than cluster scales, dynamic structure formation was intense and one speaks of non-linear scales. This distinction, accepted by astrophysicists today, will perhaps be put into question if observations reveal more important structure at large scales.

We have a good observational understanding of the non-linear distribution of galaxies. Unfortunately, we are incapable of following the complex and violent unfolding of the dynamic (i.e. non-linear) processes that operate on these scales. On large scales this is, at least according to the conventional scenarios, in principle much easier. However, it is the observations that are still incapable of clearly revealing the structure at these scales. And the little we know hardly appears to be in accord with the predictions of our models.

Let us note that the scale of 8 Mpc appears to play a privileged role. Astronomers believe they know the actual distribution of matter fairly well at this scale. The dynamicists believe, perhaps incorrectly, that they understand how dynamic processes operate. Naturally this scale is the first to be used to test the various scenarios. It plays the role of a normalisation scale.

1.2.4   Correlations and statistics

In order to compare the distribution of galaxies with the predictions of a model, or even to simply describe it quantitatively, one needs a statistical indicator. The simplest and most frequently used is the so called correlation function (CF)*. For a given distribution of galaxies this expresses the probability that one galaxy is situated near another, in other words that there is clustering.

The probability that a galaxy is at point 1 and another at point 2 may be written

$$P_{12} = n^2(1 + \xi_{12}),$$

where n is the density, or number of galaxies per unit volume, of the sample of galaxies. If the galaxies were distributed completely randomly and uniformly, this probability would be simply n^2. The quantity ξ_{12}, or simply

ξ, is the two point correlation function and indicates that the distribution is not uniformly random, but that the galaxies have a tendency to cluster. The overall homogeneity of the universe implies that ξ only depends on the distance between the two points considered. $\xi(r)$ measures the mean number of neighbouring galaxies per unit volume at distance r to a given galaxy in excess of a uniformly random distribution.

The function $\xi(r)$, measured from a number of galaxy samples, appears to take on the universal value

$$\xi(r) = \left(\frac{r}{r_0}\right)^{-\gamma}$$

where $\gamma \approx 1.7$ and r_0 is about 5 Mpc. This is valid for distances between 0.5 and 10 Mpc.

The two point correlation functions are only a first approach. One can equally define 3, 4,... point correlation functions, although these are more difficult to measure. Other indicators such as the void and count probabilities, multidimensional analysis etc. are used too. These analyses are intricate and the results difficult to interpret because we are unable to predict the values according to whichever model we take. Only numerical simulations can give us some indication of the model predictions in the non-linear regime, and even then it is difficult to extract reliable information from these.

One of the interesting results of these measurements is that they show the existence of a scaling invariance in the matter distribution of the universe. This obeys similar laws whether one considers large or small scales. This property needs to be confirmed and interpreted. A priori it is unexpected in the majority of galaxy formation scenarios. Without doubt it is a trail that will help us understand cosmogony better.

At linear scales observations are scarce and disparate, with the various bias and selection effects poorly understood. It is difficult to measure accurately the indicators mentioned above. Nevertheless it appears that matter is highly organised at large scales; we have already spoken of filaments, sheets and large voids, etc. The problem is to know how to characterise the presence of these structures, and their properties in ways other than their simply visual appearance.

One of the problems that this poses is whether these inhomogeneities extend indefinitely with increasing scales. Finally, are they compatible with the hypothesis we made at the beginning that the dynamics and the geometry of the universe can be calculated using a constant mass density? It is worthwhile remembering that each time new observations have become available, they have indicated inhomogeneities at these newly investigated scales. Will this process stop? If so at what scale? It is certainly too early to say.

We should remember that at large distances we observe more quasars than galaxies. It would thus be tempting to study their distribution. This approach has not so far been very fruitful for two reasons. The first is that the nature of these objects is still a mystery. We know neither the relationship between quasars and galaxies nor, more generally, that between quasars and other forms of matter. Thus even if we knew the distribution of quasars well, we would not know what significance to accord it. We would not be able to deduce from it the distribution of matter in the universe. (Some models predict that there should be quasars precisely in parts of the universe where there is no matter.)

On the other hand, the survey of quasars is still very incomplete. As with the galaxies, we still do not have at our disposal complete catalogues in large regions of the sky. For this reason it is difficult to use them to study in any quantitative way the distribution of matter.

1.3 Distances and velocities in the universe

1.3.1 Methods and how they complement each other

To investigate and understand the universe one needs to know the distribution of objects that inhabit it. There is no difficulty in giving the coordinates of a star or galaxy on the celestial sphere, but how can you tell whether one galaxy is more distant than another? And how can we measure its distance? This is a fundamental problem in astronomy and cosmology. A step by step approach has to be adopted, and the question is not completely resolved.

Various methods are available, applicable at different and increasing distances. One is the method of parallaxes*, which is very direct but only applicable to nearby objects. Ultimately the only method that can be applied to distant objects is to use redshifts*, although this method is indirect and subject to great uncertainties.

The method of parallaxes can be used to determine distances to nearby stars. In principle it is straightforward, and uses triangulation. Unfortunately it only applies to nearby stars within our own galaxy, and cannot be used directly in cosmology.

These nearby stars nevertheless play an important role in the estimation of greater distances. In fact it is easy to measure the apparent brightness of a star. If the distance is known (by the method of parallaxes), we can deduce its absolute brightness. The star can thus play the role of a primary indicator; if one can recognise amongst the stars in a nearby galaxy one

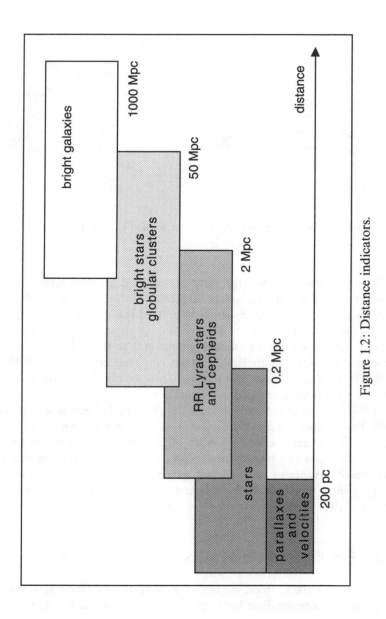

Figure 1.2: Distance indicators.

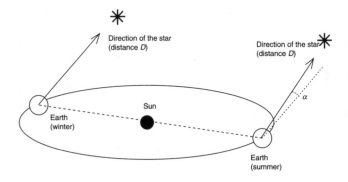

Figure 1.3: One measures the distance D of a star by its parallax α.

which is identical, i.e. has the same colour and spectral type, one can reasonably assume that its absolute brightness is also the same. In this case it is the measurement of its apparent brightness that enables us to measure its distance. In particular the stars called Cepheids have a remarkable property that makes them particularly amenable to this role. The distance of distant Cepheids is established by comparing their brightness with that of the nearby Cepheids in our own Galaxy. Thus we can deduce the distances of galaxies containing recognisable Cepheid stars.

In this way what astronomers call the distance ladder* is set up. One establishes by direct means the distances of primary indicators, such as the Cepheids. These allow one to establish the distances to nearby galaxies. In these galaxies one can identify new classes of object that are brighter, and therefore observable in more distant galaxies (see figure 1.3.1). These in turn play the role of secondary indicators. But everything begins with the method of parallaxes.

1.3.2   Parallaxes of nearby stars

Triangulation is the simplest method of measuring a distance, as any sailor will be aware. Astronomers have applied this method for a long time (see figure 1.3).

The principle is simple. It involves observing the same object, a star, from two different points P_1 and P_2 separated by a distance B, called the base. From each of these points one can measure the angles θ_1 and θ_2 that the object makes with the horizontal. Elementary trigonometry yields the value of the difference $\alpha = \theta_1 - \theta_2$, which is assumed to be small. Thus

$$\alpha = \frac{B \sin \theta_1}{D}.$$

The accuracy of this method increases with the length of the base, but no distance on Earth is sufficient to allow the triangulation of a star. However, astronomers are able to utilise the motion of the Earth around the Sun. By making observations at six month intervals they effectively are placing themselves at positions separated by the diameter of the Earth's orbit around the Sun, which thus has the role of base. By definition, the parallax of a star is the angular difference α measured with this base.

This method is so fundamental that astronomers have used it as a basis for defining their unit of distance: the parsec, (pc), is the distance of a star whose parallax is one second of arc ($''$). The nearest star is at 1.3 pc. In other words the Earth's orbit subtends an angle of $1/1.3 = 0.7''$ at the star. The distances to several thousand stars have been measured using parallaxes, out to 30 or so parsecs. The satellite Hipparcos, launched several years ago, has provided data that will extend and improve these measurements.

1.3.3 The distance ladder

Unfortunately galaxies, the things of interest to the cosmologist, are too distant for the application of this approach. The most immediate method for estimating the distance of a galaxy is to compare its brightness (or its size) with that of a nearer galaxy, or rather to compare the brightness of a star contained within the galaxy with that of a star of similar type at a known distance. One calls such a star a standard star. One simply uses the relation

absolute luminosity = apparent luminosity × distance.

The method works better if one uses a gas cloud or star cluster, which is brighter and therefore easier to measure. Rather than the brightness, one can also use dimensions as a distance indicator, according to the formula

real diameter = angular diameter × distance.

We should note in passing that these two formulae use, following geo-metrical (Euclidean) notions, the same quantity that we have called distance, whereas this is no longer valid in cosmology for distant objects. It will be necessary to define two distinct distances, an angular distance and a bright-ness distance, as we shall describe in subsection 2.2.6. For nearby galaxies there is no practical distinction between these, and one can simply talk of the distance. But this is not true for the most distant galaxies or quasars.

Astronomers thus use a chain of such references: variable stars (Cepheids) or giant stars, gas clouds (HII regions), star clusters or even galaxies themselves. But these estimates are imprecise and leave an uncertainty of a factor of up to 2 in the determination of the distances of nearby galaxies. These uncertainties have repercussions all the way up the distance scale. Today there is no precise method to directly estimate the distance of a very distant galaxy or object. Usually astronomers use a very indirect method, based on the law for the expansion of the universe described below and on measurement of redshifts.

1.3.4 The expansion of the universe and redshifts

Astronomical objects such as galaxies and quasars are more or less fixed within the framework of the universe, but this framework is not rigid or static. It is in expansion, and this expansion carries the galaxies and quasars along with it. They appear to recede from us. Since this is not a material framework it is impossible to directly record its evolution. It is revealed by the motion of the galaxies. The fundamental property of the cosmos, the expansion of the universe, is thus an observational fact. The recognition of this expansion as a property of the underlying cosmological geometry, and not simply as a property of the galaxies manifesting it, was the crucial step forward in twentieth century cosmology (see figure 1.4).

The American astronomer, Edwin Hubble, holds the credit for having discovered, in the first two decades of this century, that galaxies recede from us. This law now bears his name. Apart from the closest neighbour galaxies, there is not one that is coming towards us.

This expansion is best interpreted within the framework of relativity. It was nevertheless established within the framework of non-relativistic physics and, for reasons of simplicity, we shall present it in this way. However, one should bear in mind that this expansion can only be correctly described within the relativistic framework given in chapters 2 and 3.

Observations allow us to measure the velocity of a galaxy relative to us, or at least the radial component of this velocity, by which we mean the recessional velocity (or possibly the velocity of approach) in the direction of the line joining us to that galaxy. The laws of propagation of radiation state that radiation, for example visible light or radio waves, is received at a different frequency from the emitted frequency if the source is in motion. The observed frequency, $f_{observed}$, can be measured spectroscopically. We can also find the emitted frequency, $f_{emitted}$, which results from some physical process in the observed galaxy, since the laws of physics, in the main atomic

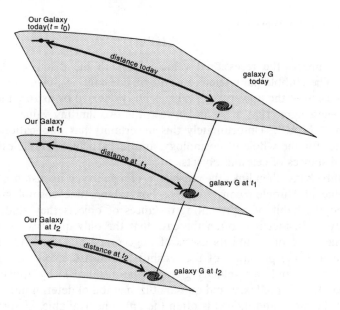

Figure 1.4: The expansion of the universe.

physics, allow us to calculate it. Usually it is a spectral line that one measures, that is the component of the radiation emitted (or absorbed) at a well defined frequency.

The redshift

The fundamental quantity is the redshift defined as

$$z = \frac{f_{\text{emitted}}}{f_{\text{observed}}} - 1.$$

The Doppler relation states that it is linked to the radial velocity according to the formula

$$z = \left(\frac{1 + v/c}{1 - v/c} \right)^{1/2} - 1,$$

which reduces to $z = v/c$ for velocities much less than the speed of light, c. The sign of z indicates whether the galaxy is moving away from or towards us.

Hubble discovered not only that all galaxies are moving away from us but that their velocities are proportional to their distance from us. This is expressed in Hubble's law,

$$v = H_0 D,$$

which connects the recessional velocity, v, with the distance, D, of the galaxy. The Hubble parameter*, or Hubble constant, H_0, measures the rate of expansion of the universe. Its value is hotly debated by astronomers, who nevertheless agree that it lies in between the two limiting values of 50 and $100 \, \mathrm{km \, s^{-1} \, Mpc^{-1}}$. Unfortunately this uncertainty has implications, as we shall see, for the whole of cosmology, and in particular for the estimation of the distances of celestial objects.

Hubble's law then links the redshift, z, of a galaxy to its distance through the value of Hubble's constant, H_0. For a long time cosmologists have adopted the habit of representing distances of objects they study by the quantity z. In fact it is often the case that the only quantity available to us is the value of z, and its use is of huge practical value. According to the classical interpretation of the redshift, the measurement of z gives us the velocity v and this velocity gives us the distance D by application of Hubble's law. For galaxies and quasars this method of determining distances is used the most, and indeed is often the only one available. Unfortunately, it is not perfect for several reasons.

In the first place in order to convert z to distance one needs the value of H_0, and this is poorly determined. It would be necessary in fact to 'calibrate' Hubble's law, or in other words measure both distances and velocities for as many far off objects as possible. In this process the high errors on distance estimates of galaxies (the problem of the distance ladder) have consequences for the estimated value of H_0. The value H_0 appears to be in between 50 and $100 \, \mathrm{km \, s^{-1} \, Mpc^{-1}}$. It is convenient to introduce the dimensionless parameter $h = H_0 / 100 \, \mathrm{km \, s^{-1} \, Mpc^{-1}}$ and rewrite Hubble's law in the form

$$D = \frac{v}{100h}.$$

In other words the uncertainty in H_0 appears in the unit $h^{-1} \mathrm{Mpc}$ used to express distances and allows us to write down rigorous expressions.

The second problem is that the velocity of a galaxy is not simply its expansion velocity, given by Hubble's law. Either because the galaxy experiences the attraction of another galaxy or cluster of galaxies, or because it has kept an initial velocity, it can have an additional velocity, called its peculiar motion, as opposed to its 'collective' velocity of expansion. Because of this unknown contribution to the expansion velocity, Hubble's law is ineffective for applications demanding precision. Thus for example we know that nearby neighbouring galaxies as well as our own are pulled towards Virgo, the cluster nearest us. To estimate distances of nearby galaxies from

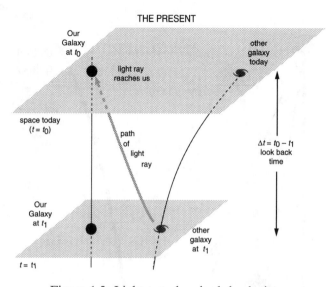

Figure 1.5: Light travel or look-back time.

redshifts one needs to take into account the fact that this 'fall-in towards Virgo' is added to the expansion velocity. Several models have been proposed to correct for this effect so that true distances may be estimated, but none is fully satisfactory.

1.3.5 Out into space, back in time

When it comes to far distant objects, beyond one hundred or so Mpc, there is a still more fundamental problem. Simply put, we can say that the most distant galaxies are seen not as they are today but as they were a time Δt in the past. Δt, the 'look-back' time, is the time taken by light emitted by the galaxy to reach us (of the order of D/c, where D is the distance to the galaxy) (see figure 1.5). For a galaxy a thousand million light years away this light was emitted about one thousand million years ago. But at that epoch the universe was younger by an amount Δt and its rate of expansion different. This light has the characteristics it had when it was emitted. Thus it is necessary to use the value of Hubble's constant then rather than its value today in Hubble's law. Hubble's law must in fact be formulated and interpreted relativistically, which is the only natural way of viewing things.

We shall see that it is possible in principle to give the distance and the time of any cosmic event if we know the redshift, z. To do this, however,

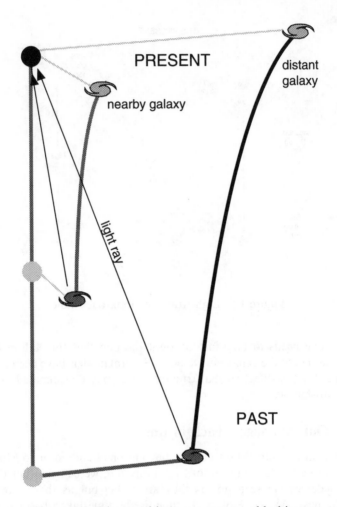

Figure 1.6: Distant objects are seen as old objects.

we need to know the law of evolution of H as a function of time, to know the dynamical parameters of the universe. For this one must adopt a cosmological model.

Finally, in order to avoid all ambiguity related to the Hubble constant and the cosmological model, astronomers fix galaxies and other objects by their redshift, z, which has a couple of advantages. On the one hand it is the directly measured quantity; errors arising from the distance ladder do not intervene. On the other hand, use of z does not presuppose a choice of any particular cosmological model.

In addition the use of z allows us for the first time to consider the nature of the intervals, half temporal, half spatial, that separate us from other galaxies we see. In fact, because of the propagation time of light, we receive light emitted by the observed galaxy a long time ago, often several thousand million years ago. The higher the redshift z the younger the galaxy must be (see figure 1.6).

Thus z, which is directly measured, fixes the distance both in space and time. Higher values of z correspond to events in the more distant past. Mostly the redshifts of galaxies are less than 1, although one galaxy has been observed with a redshift of 3.8. Quasars, which are the most distant observed objects, have been observed with redshifts greater than 4. The oldest event observable corresponds to a redshift of around 1500, whereas the very beginnings of the universe go back still further. The first instant of this expansion, in this system, corresponds to an infinite value of z.

The only way of accounting for these effects correctly, however, is to adopt a relativistic description of the universe.

2

The relativistic universe

2.1 The theories of relativity

The observation that the universe expands forces us to revise our notion of space and time. This is a crucial conceptual jump. It requires one to abandon the idea of the rigid and absolute space of Newtonian physics and accept a space that dilates! We shall see how the general theory of relativity not only permits but even predicts such an expansion. General relativity allows us to consider an evolving geometry, thus flatly contradicting the idea of absolute and rigid space central to Newtonian physics. It strips the notions of time and space of their absolute and autonomous nature, and imposes in their place the new notion of spacetime*. This spacetime necessarily has a complicated geometric structure. Such is the framework within which cosmology must be described.

2.1.1 Special relativity

Both theories of relativity are due to Einstein (the second encompassing the first). Special relativity describes a 'virtual' spacetime, free from gravitation. It furnishes an understanding of a whole number of properties related to causality and to the concept of spacetime. Nevertheless general relativity is indispensable to any cosmological description as it alone allows a coherent treatment of gravitational phenomena taking place on the scale of the cosmos.

The essential postulate of special relativity is that the speed of light is constant (in a vacuum) no matter what the state of motion of the source or the observer. No object we observe is contemporaneous with us; light from the sun takes about 8 minutes to reach us, from stars several years, and from distant galaxies several thousand million years. This time is called the 'light travel time'* (see subsection 1.3.5). When we look back in space we also look back into the past and observe different sections of the universe; the more distant they are the older they are. A far distant galaxy is observed as it was when very young, shortly after its birth a few thousand million years ago. On the other hand, a nearby galaxy appears to us having already evolved over a period of thousands of millions of years. Evidently these objects will be very different, as different as an old man from an infant.

What separates us from another galaxy is just as much a spatial interval (a distance) as a time interval (a period of time). Strictly speaking it is neither one nor the other. This is one of the novelties of cosmology, and one of its difficulties.

As soon as one considers the distant universe, one has also to take into account this duality of intervals; one can no longer talk of points in space but has to introduce the notion of an event of spacetime; a galaxy there and then, another here and now. This is the reason why it is more satisfactory to characterise an observed object by its redshift, z, rather than by its distance or time separation. This is the ideal 'mixed' quantity for cosmologists. Special relativity forces this change upon us. It states that distances and time intervals are only relative, and depend on the state of the observer. The only absolute quantity is formed from this mixed interval, and combines both temporal and spatial aspects in the form of a spacetime metric.

2.1.2 The general relativistic universe

General relativity (GR) forces us to make a more radical revision of our concepts. In non-relativistic physics geometry is simple. (In Euclidean geometry, which we learn in school, if one travels straight ahead one never returns to the point one started off from etc.) Thus any model of the universe that does not take general relativity into account is necessarily very simple. Space can only have this simple geometry which does not evolve in time. Any discussion of the universe is thus rather trivial. Nor is it possible to give a fully coherent description of the universe within this framework.

In contrast, general relativity offers a rich diversity of possible geometries. Euclidean geometry is just one special case. These geometries define, although not completely, the spatial extent of the universe, whether it is finite or infinite, the laws for the propagation of light as well as other properties which are trivial in ordinary geometry. Geometry is not static but evolves in the course of time. The universe not only has a structure, it also has an history.

It is in fact better to talk simultaneously of the geometry of space and its evolution as two aspects of an 'extended' geometry describing not just space as we know it, with three dimensions (height, width and depth), but a four dimensional spacetime where time plays the role of the fourth dimension. The theory treats the geometric and the temporal aspects of the cosmos in a similar manner. In this way the fact that absolute space and time do not exist is built into the theory.

At least three conceptual novelties due to relativity must be built into cosmology. They might complicate matters, but they also add a tremendous richness to the theory. They form an indispensable foundation to today's cosmology. Firstly the notion of space and time as separate entities is replaced by the notion of spacetime, an innovation of special relativity. The second is to introduce non-trivial, ie non-Euclidean, geometries. The last is to consider the evolution of this generalised geometry. One immediate difficulty arising in relativistic cosmology stems from the fact that we are all used to three dimensional geometry, but not four dimensional. When considering a straight line, a plane and 'ordinary' space we pass from one to two and finally to three dimensions. Yet the next step of going to four dimensions, even in Euclidean space, is almost impossible for us to grasp. A second difficulty derives from the unusual structural properties of this geometry, which we occasionally describe by the term 'curvature' of space or of spacetime. Mathematicians call such a generalised space a manifold*, or, in the case that interests us, a differentiable manifold or Riemannian manifold. One could say that the spacetime manifold is as complicated, and thus as rich in structure, in relation to Euclidean space in four dimensions as an arbitrary surface can be when compared with the plane. Luckily mathematics provides a description of such structures, although this description is local rather than global. Thus at each point the so called metric properties can be described. This allows one to calculate lengths of time and distance, or rather intervals combining the two, given the inseparability of temporal and spatial properties. The entire geometrical structure of the cosmos, and indeed its own evolution is expressed by the metric* ds^2.

Metric theories state that spacetime can be deformed. General relativity is the most attractive and the best observationally supported of these theories. Its originality lies in the way in which it predicts that the energy content of the universe determines the structure of the universe. This is why the matter content of the universe is so important for cosmological questions. Essentially the construction of a cosmological model is tantamount to solving, at least in part, the equations of general relativity, the so-called Einstein equations. The geometrical structure of the universe and the evolution of this structure can similarly be determined, though not independently of the matter contents of the universe. Since we only have partial information about the contents of the universe, we have to introduce a number of further assumptions in order to make any headway. One such assumption is homogeneity, which also goes by the name of the cosmological principle.

2.1.3   The metric of Euclidean space

Relativistic cosmology requires the use of a metric. This metric is indispensable for describing spacetime. One can understand the significance of the concept of a metric in the context of ordinary space in three dimensions and with coordinates (x, y, z). The square of the distance between two neighbouring points, separated by small quantities dx in height, dy in width and dz in depth can be written simply as

$$d\sigma^2 = dx^2 + dy^2 + dz^2. \tag{2.1}$$

$d\sigma^2$ is called the line element and the distance between two points is obtained simply by evaluating the integral of $d\sigma$. For ordinary space the integral is trivial and leads to the usual formula for the distance between two points.

We could just as well have written the metric of ordinary space in non-Cartesian coordinates, such as spherical polar or cylindrical coordinates etc. For example, in spherical polar coordinates the metric becomes

$$d\sigma^2 = dr^2 + r^2 \left(d\theta^2 + \sin^2 \theta d\phi^2\right) \tag{2.2}$$

(see figure 2.1).

Going from one form to the other simply requires a change in coordinates. Both forms describe the same (ordinary) geometry, but in different ways. Which form one uses to evaluate the distance between two points is unimportant.

The metric also allows one to calculate surface areas and volumes. For example, the volume element of size dx, dy and dz may be written

$$d^3V = dx \, dy \, dz \tag{2.3}$$

or, in spherical polar coordinates

$$d^3V = r^2 \, dr \, d\theta \, \sin \theta \, d\phi \tag{2.4}$$

The volume of a region of space is the triple integral of d^3V within the given region. For example, to calculate the volume of a sphere of radius R, the integral over the angles separates out and gives 4π, the solid angle over all directions, whilst the radial part integrates to yield $R^3/3$, which gives the volume of the sphere as $(4/3)\pi R^3$.

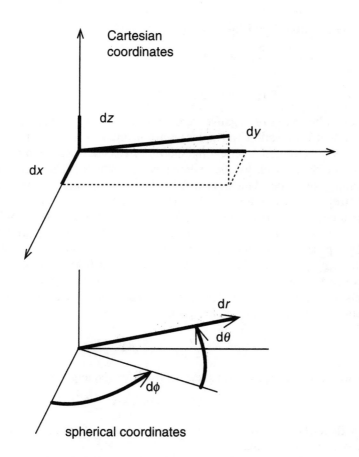

Figure 2.1: Coordinates for the Euclidean metric.

2.1.4 The metric for curved spaces

The metric for a curved space is of course more complicated. Currently used cosmological models are based on the assumption that space is homogeneous and isotropic, and we shall restrict ourselves to this case. There are mathematical theorems that tell us that there are only three types of space (locally), and these have relatively simple metrics. They can be distinguished by the value of a curvature constant, k. Thus we can talk of spherical ($k = 1$), flat ($k = 0$), or hyperbolic ($k = -1$) spaces. The metric of a homogeneous and isotropic space can always be put in the form

$$d\sigma^2 = d\chi^2 + S_k^2(\chi) \, (d\theta^2 + \sin^2\theta \, d\phi^2), \tag{2.5}$$

where the function $S_k(\chi)$ is defined for the different values of k as

$$
\begin{aligned}
S_k(\chi) &= \sinh\chi && \text{for } k = -1, \\
S_k(\chi) &= \chi && \text{for } k = 0, \\
S_k(\chi) &= \sin\chi && \text{for } k = 1.
\end{aligned}
\tag{2.6}
$$

These describe spaces of negative, zero, and positive curvature respectively.

An initial observation will guide us to an interpretation of the above. When $k = 0$ we obtain precisely the metric of Euclidean space in spherical polar coordinates given by the expression (2.2) where χ plays the role of the radial coordinate. This is evidently one of the possible homogeneous and isotropic spaces, ordinary geometry. The coordinate χ, which is called r in ordinary geometry, varies from 0 to infinity. The angular coordinate θ, analogous to the latitude or in astronomy to the declination, varies from 0 to 2π. Space with negative curvature is described, as in the case of zero curvature, by χ running from 0 to ∞. χ runs from 0 to π for space with positive curvature.

This metric is occasionally written in an equivalent form with $r = S_k(\chi)$ replacing χ as a radial coordinate so that

$$
d\sigma^2 = \frac{dr^2}{1 - kr^2} + r^2 \, d\Omega^2,
\tag{2.7}
$$

where we have used the abbreviated notation for the angular part

$$
d\Omega^2 = (d\theta^2 + \sin^2\theta \, d\phi^2).
$$

The factor $(1 - kr^2)$ for $k \neq 0$ in the denominator of the metric expresses the existence of curvature.

The r coordinate varies from 0 to infinity if $k = 0$ or -1. However for a space with positive curvature r varies from 0 to 1 for the first half of space (χ running form 0 to $\pi/2$). In order to describe the other half (χ running form $\pi/2$ to π) one has to allow r to decrease from 1 to 0. This is often a source of confusion (see figure 2.2).

2.1.5 Lengths and volumes

The value of the constant k indicates the sign of the curvature (zero if $k = 0$). As in a space without curvature, one can calculate lengths, angles, volumes etc. from the metric. Comparison with the properties of ordinary space provides one with a guide for analysing curved geometry. In order to do this we choose our position as the origin of the coordinate system with $r = 0$ (or $\chi = 0$). Thus coordinates r, θ and ϕ (or equivalently χ, θ and ϕ) can be identified with spherical polar coordinates centred on the observer.

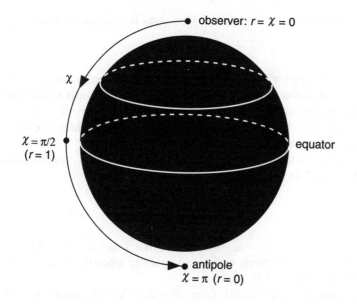

Figure 2.2: Spherical space has a finite volume. It is measured by the coordinate χ.

Consider a point of the space with radial coordinate $r = r_1$ (or equivalently $\chi = \chi_1$). We can assume without any loss of generality that the other coordinates are given by $\theta = \phi = 0$. The integral of the metric gives immediately the distance between this point and us as $\chi_1 = \text{arc}S_k(r_1)$. The function $\text{arc}S_k(r)$ is defined as the inverse of the function S_k, that is $\sin^{-1} r$ for $k = 1$, r for $k = 0$ and $\sinh^{-1} r$ for $k = -1$.

The coordinate χ is interpreted as the spatial distance between a given point and us. In spacetime we shall see that it represents, up to a multiplicative factor, the so-called proper distance. Clearly, contrary to common practice, it is more practical to use the coordinate χ than r.

The difference between this space and ordinary space appears when we calculate the circumference of a circle centred on the observer (ourselves) and passing through a point with radial coordinate r, or χ, according to the description we adopt. In fact such a circle (for which it can be assumed that θ is constant) is defined by $r = r_1$ with ϕ varying between 0 and 2π. The length of the circumference of this circle is obtained by putting

$$dr = 0,$$
$$r = r_1,$$
$$\theta = \pi/2$$

into the expression for the metric. This gives for the circumference

$$2\pi r = 2\pi\, S_k(r).$$

Explicitly $2\pi \sin \chi$ for $k = 1$, $2\pi\chi$ for $k = 0$ and $2\pi \sinh \chi$ for $k = -1$.

These calculations show that the coordinate χ represents a radial distance, that is a radius, whilst the coordinate r represents, up to a factor of 2π, a circumference. In Euclidean space the distinction, which is essentially a manifestation of the curvature of space, disappears.

Thus, according to our calculations, the ratio of the circumference of a circle to its radius is given by

$$\frac{2\pi\, S_k(\chi)}{\chi}.$$

Only in the Euclidean case does it take the value 2π. It is less than 2π when the space is positively curved, and greater when the space has negative curvature (see figure 2.3).

One deduces immediately from this that in the spherical model the distance between two points cannot exceed π, since the coordinate χ varies between 0 and π (in the same way the angle of separation of two points on the surface of a sphere is always less than π).

The effects of curvature are also manifested in calculations of volumes. The element of volume may be written

$$d^3V = d\chi\, S_k(\chi)d\theta\, \sin\theta\, d\phi = (1 - kr^2)^{-1/2}\, dr\, r^2\, d\theta \sin\theta\, d\phi.$$

One has to integrate this over the given volume.

One can thus calculate the volume of a sphere centred on the observer, which only reduces to the the usual value of $(4/3)\pi r^3$ in the case of Euclidean space ($k = 0$).

The size of the universe

These formulae can be applied to the whole universe. In Euclidean space as in hyperbolic space, the proper distance to a point can tend to infinity (since it is equal to the coordinate χ, which itself can take arbitrarily large values). It is the same for the circumference of a circle, and for the volume of a sphere. These quantities tend towards infinity with the coordinate χ. Both these types of space then have infinite volume.

In the spherical case things work out differently. The whole of the space is covered if the coordinate χ varies between 0 and π (in exactly the same way as for the latitude of a spherical surface). In this case it is fairly easy

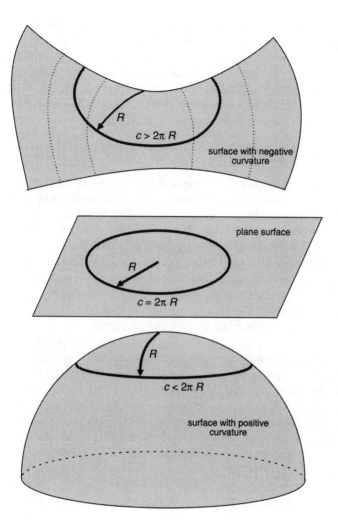

Figure 2.3: The circumference of a circle of radius R is respectively greater than, equal to, or less than $2\pi R$ according to whether the curvature of space is negative, zero, or positive.

to see that the maximum distance to a point from the observer is precisely π. The corresponding point is the most distant point from the observer. It is unique, and called the antipodal point in analogy to the sphere. At this point the r coordinate takes the value 0, in such a way that a circumference centred on the origin and passing through this point has zero length. This is not very surprising, since it reduces to a point! It is also easy to show that a sphere passing through the antipodal point and centred on the observer has volume of $2\pi^2$, and that it contains every point of space. Spherical space is finite and has a volume of $2\pi^2$. One should also note that the maximum circumference of a circle centred on the observer is 2π, and occurs when the r coordinate takes its maximum value of 1 (or χ the value of $\pi/2$). One sometimes calls this circle the equator (see figure 2.4).

2.1.6   Spacetime metrics

We have studied the metric of space. From the point of view of relativity, and cosmology, space is a spatial section (at constant time) of spacetime. To write down separate metrics for space and for time means one is working separately in space and in time. But in practice this is never possible. For example, the trajectory of a light ray evidently does not take place at either a fixed position or at a fixed time. It is rare in cosmology to face a problem that concerns only the spatial (and non-temporal) properties of spacetime. In fact, we see far distant objects after a delay of the light travel time. Even so, the analysis of the spatial properties is essential to understanding the nature of the space described by this sort of metric.

If we are to use relativity, we have to work in four dimensional space, and manipulate intervals other than simple distances and time durations. We have to define a spacetime metric. Special relativity is useful for describing a universe devoid of gravitation. Of course this is unrealistic, but its simplicity suggests that it might aid us in understanding some of the fundamental cosmological problems. Its metric may be written in a very simple form where the contributions of space and time are explicit. (In order for the units of time and space to be compatible, we multiply the time by c^2.) Thus in Euclidean* spacetime (in fact we talk of pseudo-Euclidean or Minkowskian spacetime because of the minus sign) the interval between two events separated by dx, dy, dz, in space and dt in time is given by

$$ds^2 = c^2\,dt^2 - d\sigma^2, \tag{2.8}$$

where

$$d\sigma^2 = dx^2 + dy^2 + dz^2 \tag{2.9}$$

represents the Euclidean spatial metric which we described above. Time, or rather the variable ct, plays the role (except for the fundamentally important minus sign) of a fourth dimension, expressed in Cartesian coordinates.

This metric allows a given observer to determine the distance between two points. He thus has to consider these at the same instant, and suppose that the time part, dt, separating them is zero. The spatial metric is then

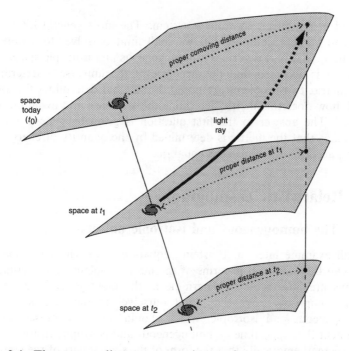

Figure 2.4: The proper distance separating us from another galaxy grows with the expansion. The separation an object is from us is measured by its proper comoving distance.

$d\sigma^2 = -ds^2$, which can be integrated between the values of the spatial coordinates of the two points. We should emphasise the point that another observer moving with respect to the first will not find the same result for this operation.

One can also calculate the time between two events that occur at the same point in space. They have the same spatial coordinates, so now the spatial part is zero, and one obtains the interval of time from $ds = dt$.

Special relativity assumes a flat space and spacetime without curvature, that is a Euclidean space and spacetime. It is for this reason that the form of the metric is simple. General relativity, on the other hand, furnishes spacetime with a much richer structure, and describes it by a much more complex metric.

2.1.7   The metric of curved spacetime

The metric expresses the properties of spacetime, but only locally. The line element involves the differences in spatial and temporal coordinates expressed in any given coordinate system. The most general form of this metric can be very complex, but we shall limit ourselves to simple cases where the spatial part describes a homogeneous and isotropic space.

Once it is accepted that the structure of the universe is described by such a metric, two problems are posed. First, what determines this metric? Second, how does this metric determine the structure and evolution of the universe? The answer to the first question is given by general relativity, which states that the metric is determined by the quantity of energy in the universe, according to Einstein's equation.

2.2   Relativistic cosmography

2.2.1   The homogeneous and isotropic universe

We shall examine later on Einstein's equation from which the spacetime metric can be calculated in terms of the matter content of spacetime. The equations are impossible to determine in the most general case because of their complexity. We shall limit ourselves to the case where space is homogeneous and isotropic, which should not be confused with the requirement that spacetime be homogeneous and isotropic. In this case one can prove mathematically that the form of the spatial part of the metric (i.e. the metric obtained by putting $dt = 0$) can be written in the form above.

If one assumes that space and time are separated in this way, and that space is homogeneous and isotropic, all possible metrics can be written in the common form

$$ds^2 = c^2\, dt^2 - R^2(t)\, d\sigma^2. \tag{2.10}$$

The spatial part of this metric is $R^2(t)\, d\sigma^2$ where the term

$$d\sigma^2 = d\chi^2 + S_k^2(\chi)\, (d\theta^2 + \sin^2\theta\, d\phi^2)$$

describes space as we indicated above in (2.5). Let us remember that it can also be put in the form given in equation (2.7),

$$\sigma^2 = \frac{dr^2}{1 - kr^2} + r^2\, d\Omega^2,$$

if one uses the coordinate $r = S_k(\chi)$.

The purely spatial properties of the universe, which can be calculated from the spatial part of the metric, involve only relations between points of the cosmos taken at the same instant t of cosmic history, or in other words simultaneous events. However, the cosmologist is more often interested in relations between events that are not simultaneous. Space is just a constant time surface (or hypersurface since it has three dimensions) of spacetime. Within this surface there is no variation in time (hence $dt = 0$). In this case the metric reduces to its spatial part $R^2(t) \, d\sigma^2$.

It should be noted that this spatial metric changes in the course of time through the term $R(t)$, which acts as a scale factor. It does not change the characteristics of the geometry, but effectively redefines the unit of length. For this reason it is called the scale factor*. The evolution of this scale factor is fundamentally important, and expresses the fact that as a result of cosmic expansion the scale of space changes with time.

The form of the spatial metric shows that at a given time t space is identical to the model space we described in subsection 2.1.4, provided one simply multiplies all lengths and distance elements by the scale factor $R(t)$. Thus the structure is the same but all lengths must be rescaled. The distance between any two fixed points in space is proportional to $R(t)$; volumes are proportional to $R^3(t)$. Hence if the function $R(t)$ increases, space 'inflates' or 'expands' with time.

2.2.2 Expansion

The preceding calculations are in some sense purely theoretical because they apply to the geometry of space frozen at some instant t. Yet we do not have access to such a space since our observations are of objects in our past. If we observe a galaxy we do not see it contemporaneously, but as it was a time Δt in the past (at the retarded time). Thus we have to consider an event separated from us in both space and time. How can such an event be fixed? How can it be described? Since the event is not contemporaneous with us, it is a somewhat delicate question to talk of its distance from us.

We can try to define a proper distance. However a proper distance is only defined for two simultaneous events, and one has to choose some instant of time to evaluate it; either the time of emission t_1, or the time of observation t_0. These two choices define two proper distances, $d_p(t_1)$ and $d_p(t_0)$. So which is the better choice?

Comoving quantities

It seems more rational to fix objects by their position in comoving space today, and this in general is the choice astronomers make. Take for example a galaxy at r_1 or χ_1, whose light emitted at time t_1 reaches us today. We mentally reconstruct the position that this galaxy would occupy today, assuming that it has followed the overall cosmic expansion. The corresponding comoving proper distance is $d_p = R_0 \chi_1$, where $R_0 = R(t_0)$ is the value of the scale factor today. If we had estimated the proper distance at time t_1

we would have obtained $R(t_1)\chi$. All cosmic distances increase with cosmic time, whilst the coordinates of the extremities remain constant (see figure 2.4).

Objects such as galaxies which are in free fall are by definition stationary with respect to the geometry, which is itself expanding (there is no interaction to move them in this geometry). In this case their r and χ coordinates do not change with time even though the distance between the objects does increase as a result of the expansion. These coordinates are called comoving, and are special in this respect. Furthermore the time coordinate used is called the cosmic time and corresponds to the proper time recorded by an observer in free fall.

More generally, the proper distance, estimated at a given instant of time t, between two objects with coordinates r_1, θ_1, ϕ_1 and r_2, θ_2, ϕ_2 say, may be written

$$d_p(t) = R(t)f(r_1, \theta_1, \phi_1, r_2, \theta_2, \phi_2) = f_{12},$$

where f_{12} is a function of the comoving coordinates whose value is constant in time. This distance varies in time as $R(t)$. At the present time t_0 it defines the comoving (proper) distance between the two points,

$$D_{\text{comoving}} = R_0 f_{12},$$

which is the distance separating the two points today, assuming that they have exactly followed the law of cosmic expansion. At any instant t the proper distance between two points (fixed in the geometry), can be obtained as a function of their comoving distance according to the formula

$$d_p(t) = \frac{R(t)D_{\text{comoving}}}{R_0}. \tag{2.11}$$

In the same way a volume delimited by points that have constant comoving coordinates varies as $R^3(t)$. Thus one can define a comoving volume, which obeys the relation

$$V(t) = \frac{V_{\text{comoving}}R^3(t)}{R_0^3}.$$

The notion of a comoving quantity is very useful in cosmology since it allows one to work with a constant quantity, such as distance, volume and density, whilst the true quantity varies with time.

2.2.3   The relativistic universe

Up to now we have supposed, and this is the fundamental assumption of the big bang, that the universe is homogeneous and isotropic. Consequently it can be described by one of the three types of metric mentioned earlier. The parameters k and $R(t)$ are still for the moment undetermined. General relativity tells us how the values of these parameters can be obtained. However, even without knowing their values, one can still analyse certain characteristics of the models.

For the cosmologist a purely spatial analysis is of limited interest since all information comes via the intermediary of radiation (today electromagnetic, tomorrow possibly gravitational). Owing to its finite speed of propagation, objects that we see now are situated in the past and it is better to talk about events corresponding to the emission of the radiation that reaches us today. The more distant the objects, the further back in the past they existed. Nor does the light that reaches us propagate in a fixed space, for the properties of space evolve as a result of the overall expansion. These effects are perfectly accounted for by general relativity, but hardly in an intuitive way. Thus it is necessary to study the trajectories and propagation of radiation.

Special relativity states that radiation always propagates at the speed of light. This is simply expressed by the relation $ds = 0$. This relation holds in general relativity too: light paths correspond to curves for which $ds = 0$, and are called null geodesics*. In fact the integral of the metric along such curves is by definition zero, since $ds = 0$ along the entire length of the curve. (Although $ds = 0$, the spatial length is not zero). From now on we shall stress the part-spatial, part-temporal nature of the trajectories. Along the path of the photon, neither the interval of time, dt, nor the space interval $R(t)d\sigma$ is zero. The interval is thus really mixed, and one can anticipate that it will not be useful to characterise it either by a distance or a time interval. It is the redshift, which has this mixed character, that is best adapted to describe it.

2.2.4 The equation for light rays

When we receive light today from a galaxy or quasar, what we observe are events that took place a long time ago. We can study these events by using a metric. We can, without loss of generality, choose the origin of our coordinate system ($r = \chi = 0$) to be at our Galaxy so that the light rays of interest to us follow radial paths. Thus we shall not need to introduce angular intervals and our analysis will involve only two variables (t and r, or χ) thus considerably simplifying the problem without losing generality. Our time coordinate takes the value t_0, corresponding to the present age of the universe.

Light reaches us radially, so that θ and ϕ are constant. Without any loss of generality we can put them equal to zero. In space the equation of this ray takes the form $\theta = \phi = 0$. The event of interest to us thus has angular coordinates $\theta = 0$ and $\phi = 0$. We shall call r_1 (or χ_1) its radial and t_1 its time coordinate.

To establish a relation between our coordinates and the emitting object's, we integrate the equation for a light ray (see figure 2.5):

$$\frac{c\,dt}{R(t)} = d\chi,$$

which yields

$$\chi = c \int_{t_1}^{t_0} \frac{dt}{R(t)}. \tag{2.12}$$

In order to evaluate the integral on the right hand side we need to know the function $R(t)$. For the moment we shall simply leave it in this form. Occasionally it is useful to introduce a new time variable, the conformal time η defined by

$$d\eta = \frac{dt}{R(t)},$$

so that the above relation can simply be written as

$$\chi = c\,(\eta_0 - \eta_1),$$

where η_0 and η_1 denote the value of the conformal time in our epoch and at the time of emission respectively.

This fundamental relation relates the time at which light is emitted to the spatial coordinate of the source, and is in some sense a generalisation of the 'ordinary' relation $R/t = c$, expressing the constancy of the speed of light. Here this constancy is expressed for a curved and expanding space.

2.2.5   The redshift

The redshift of spectral lines, which in the first analysis we interpreted as a Doppler shift due to the velocity of expansion, is better understood within a general relativistic framework. To see this, consider light of period T (and thus of frequency $v = 1/T$) emitted by a source with spatial coordinate χ.

According to the definition of period (which is in general determined by the laws of atomic or nuclear physics), the source emits two successive maxima at times t_1 and $t_1 + T_{\text{emitted}}$. Note that T_{emitted} represents the period of the light. We shall denote the period at reception, which is not necessarily the same, by T_{received}. Thus, with these definitions, we receive the two successive maxima of the light at times t_0 and $t_0 + T_{\text{received}}$ (see figure 2.6).

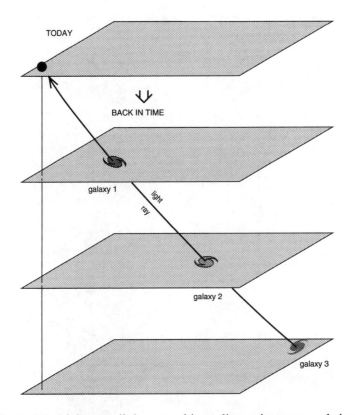

Figure 2.5: Light rays link us to objects distant in space and time.

Equation (2.12), applied to both light rays corresponding to the first and the second wave maxima, yields the relation between these time intervals. The scale factor, $R(t)$, evolves on a cosmological scale of the order of thousands of millions of years and so remains to all intents and purposes constant for time intervals comparable with the period of the light. Hence one can pull out the factor in the integrand, and finally obtain the relation

$$R(t_1)T_{\text{received}} = R(t_0)T_{\text{emitted}}$$

and further

$$\frac{T_{\text{received}}}{T_{\text{emitted}}} = \frac{\lambda_{\text{emitted}}}{\lambda_{\text{received}}} = \frac{R_0}{R} = 1 + z \tag{2.13}$$

Here z is the redshift. What is evidently important about this relation is the fact that it links z to the value of the scale factor, R, at the time of emission of the light.

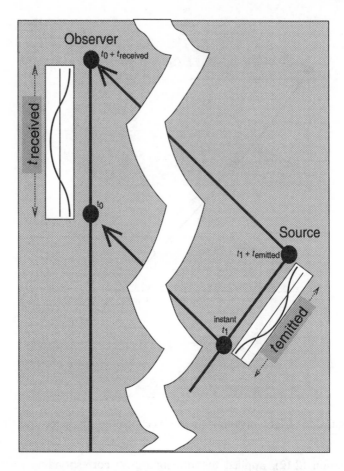

Figure 2.6: To calculate the change in the period of radiation, consider two light rays corresponding to two successive maxima of the wave emitted by a source.

There are many consequences that follow from this. In the first place, z is a measurable quantity. Moreover, it has an immediate cosmological interpretation since it gives the value of the scale factor at the time of emission of the light. Thus if we knew, for example, when the light was emitted, which unfortunately is practically never the case, we would be able to reconstruct the function $R(t)$.

As we have already noted, the quantity z is neither purely spatial or temporal in character, since it is evaluated along the trajectory of the light ray. It is ideal for measuring the interval separating us from a cosmic event linked to us by such a light ray. z is the fundamental quantity for cosmologists; it can be obtained directly from observation and is unambiguous. In contrast, the notions of distance and time intervals pose serious difficulties both conceptually and practically. If, for example, we ignore the possible observational difficulties, the redshift of a quasar is easy to determine. Determining its distance or the time at which we observe it is altogether different. (Indeed even the notion of distance needs to be clarified.)

Nearby objects have very small redshifts. $z = 1$ corresponds to a distant galaxy. Only quasars and the occasional galaxy are known with redshifts more than 3. The furthest quasar today, which is also the most distant object known to date, has a redshift of around 5, although objects of higher redshift could be discovered in the near future. Thus this number represents the present day limit on the universe of observable objects.

This is not to say that the universe, or the observable universe, stops there. However, the objects beyond are too faint to be identified. Their light does contribute however to the background radiation. In each wavelength band we observe various sources of background radiation whose precise origin, although apparently cosmological, is difficult to ascertain. The cosmic millimetre background radiation is one special case: we believe it to have been emitted, or more exactly scattered, by matter not yet structured into galaxies, at a redshift around 1500. This number represents, then, the limit of the observable universe. Astronomers have strong grounds for thinking that this will not be extended for a long time because beyond this limit the universe was opaque to electromagnetic waves. No radiation emitted beyond this limit could have reached us, except perhaps in the form of gravitational waves. But we still do not know how to detect gravitational waves. Note that this 'beyond' is a limit both in space and time.

2.2.6 Cosmological distances

To any given galaxy one can associate a proper distance, $\chi R(t)$, which varies with cosmic time. However, it is more practical to use the comoving distance, $\chi R(t_0)$, evaluated in the space of the present epoch. Unfortunately, none of these quantities is directly measurable. Moreover, since the galaxy would appear to us as it was in the past, at the time when it emitted the light, measurements can not yield the proper distance, whether comoving or not, but only a quantity that relates two events at different times. For this reason cosmologists define several different quantities, in principle measurable, corresponding to specific processes of measurement.

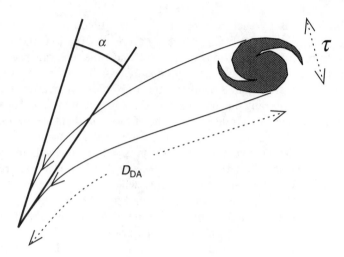

Figure 2.7: The angular diameter distance, d_{DA}

Angular diameter distance

In simple Euclidean space, it is straightforward to estimate the distance to an object of known size τ. This could be its diameter. If this object subtends an angle α, which we assume to be small, the laws of Euclidean geometry tell us that the distance, D, is given by $D = \tau/\alpha$. But this is only true in Euclidean space. In cosmology two effects modify this simple relation. The first comes from the fact that the geometry of space is not necessarily Euclidean. It could be spherical or hyperbolic. The second comes from the fact that photons do not propagate in a static space. Owing to the expansion of the universe, the geometrical properties of space vary. At each moment along its trajectory the photon passes through a space with a different scale factor, which grows in time. The only way to take this effect into account is to work with a spacetime metric.

What are the coordinates of a galaxy whose real diameter τ is assumed known (how we know this is another question), and which subtends a small angle α at the observer? Let us write down the equations for two light rays emitted at the two extremities of the galaxy's diameter, assuming that we have radial coordinate $r = 0$. These two extreme points have the same radial coordinates, $r = r_1 = r_2$, and the same angle ϕ. However, they correspond to different values, θ_1 and θ_2, of the coordinate θ. Both light rays are radial, and both have $\phi = 0$. The first describes a geodesic $\theta = \theta_1$, and the other $\theta = \theta_2$. As the observer we find that $\alpha = \theta_1 - \theta_2$ (see figure 2.7).

From the point of view of the source, the (true) distance separating the two extremal points of the galaxy can be obtained by integrating the metric between their coordinates. By definition, this gives one the angular diameter distance of the galaxy, as

$$d_{DA} = R(T_{emitted})r_1 = \frac{r_1 R_0}{1+z} \qquad (2.14)$$

d_{DA} is a directly measurable quantity. However, by measuring it alone we can neither obtain the spatial coordinates nor the redshift of the observed galaxy. In the same way measurement of z yields neither the radial coordinate value nor the distance. Nevertheless, a given specific cosmological model (fixed by the value of its parameters) provides precise relations between r, z and d_{DA}.

Conversely, simultaneous measurements of d_{DA} and z for several objects allows one, in principle at least, to determine the cosmological model that best fits the universe.

In passing we note that this formula tells us that an object of dimension τ subtends at the observer an angle of τ/d_{DA}, and furthermore, an object of surface area S subtends a solid angle of S/d_{DA}^2.

Luminosity distance

A different form of reasoning leads us to define another type of distance. In classical geometry, knowing the intrinsic or absolute luminosity, L, of the source and measuring its apparent luminosity, F, one can estimate its distance using the formula $F = L/4\pi D^2$. Thus we can define a luminosity distance through this formula.

To link this distance to the other parameters, it is necessary to take into account, as in the previous calculation, the various geometrical factors. But we also have to consider the conservation of energy of the light beam. Calculation of the power received by the detector yields the luminosity distance

$$d_L = R(t_0)\, r_1\, (1+z) = \frac{R^2\, (t_0)\, r_1}{R(T_{emitted})}$$

2.3 Causality and horizons

Distances are metrical properties, in the sense that they concern numerical measurements that are expressed by numbers. On the other hand, causal relations between events are not expressed numerically, but rather by logical relations. Are two objects, or two events, causally linked or not?

Although the notion of causal relation is fundamental, it is nevertheless somewhat vague. All physics depends on the notion of interaction. Usually the state of a system changes as a result of some interaction. An interaction always has some cause. In order for this cause to act it is necessary for these interactions to propagate from the cause (source) to the effect.

No interaction can propagate faster than the speed of light, c. In other words a ray of light is the fastest possible interaction in the universe. One question that often occurs in cosmology is whether or not two points in space, that is two spatial positions, are causally connected. The question can be formulated as follows. Is there a light ray, emitted some time after the beginning of the universe from one point, that could have reached the other point? This essentially is the problem of the causal structure of the universe.

One can fix the two points in question by their comoving spatial co-ordinates. We choose our own position to coincide with the first point with $r = \chi = 0$. Without any loss of generality, we can take the angular coordinates of the second point to be zero. Its position is thus fixed by the radial coordinate r (or χ). The equation of a light ray emitted at t_1 from point r_1 and reaching us at time t_0 can be obtained from equation (2.12).

Consider this in the context of the evolution of the universe. As we shall see later, there are two classes of model universes: those in which the present age of the universe, or at least its present phase of expansion, is finite, and those in which this age is infinite. For the first class of models, called big bang models, one can describe the entire history from some initial instant, which is chosen as time zero. On the other hand, it is impossible to talk of an initial time in models of infinite age. Let us look at the case of the big bang models. Since their history starts at time $t = 0$, we have the constraint $t_1 > t_{minimum} = 0$. Let us use this constraint in the equation above. Does it provide a constraint on the spatial coordinate?

Everything depends on whether the integral

$$\int_0^{t_0} \frac{dt}{R(t)}$$

converges or not. If it does, one can define the value

$$\chi_H(t_0) = c \int_0^{t_0} \frac{dt}{R(t)}$$

and the inequality above may be written

$$\chi_1 < \chi_H(t_0).$$

In other words, there exists in space a fictitious limit which one calls a horizon. Any event which takes place, or took place, at a point beyond this

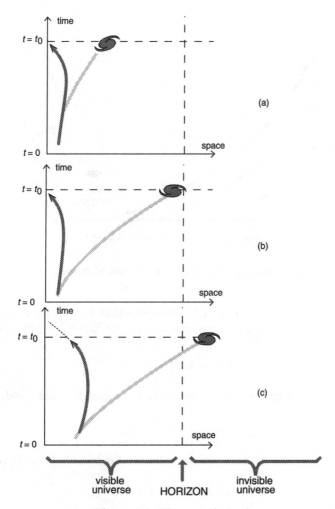

Figure 2.8: The cosmic horizon.

horizon, that is to say whose radial coordinate is greater than $\chi_H(t_0)$, could not yet have been observed by us. The light signal, or any other possible signal that might have been emitted, has not had the time since its emission to cover the distance separating us from the source (see figure 2.8).

The existence of horizons, in general relativity, is more generally applicable outside the domain of the simple model universes we have described here, and indeed of cosmology itself. There are other types of horizon (called

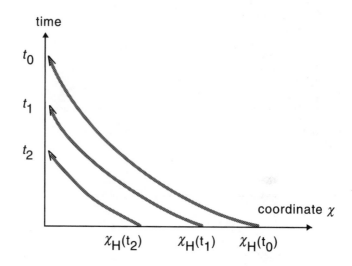

Figure 2.9: The cosmic horizon grows with time.

event horizons) which arise in relation to events that we shall never see.

For most big bang models, the expansion law for $R(t)$ is close to a power law:

$$R(t) = R_0 \left(\frac{t}{t_0}\right)^{\alpha}.$$

In this case the horizon can be immediately calculated and yields the coordinate

$$\chi_{\mathrm{H}}(t_0) = \frac{ct_0}{(1 - \alpha)R(t_0)}.$$

The proper comoving distance corresponding to the horizon is easily calculated to be

$$d_{\mathrm{H}} = R(t_0)\chi_{\mathrm{H}}(t_0),$$

which for the models obeying a power law expansion yields

$$d_{\mathrm{H}}(t_0) = \frac{ct_0}{1 - \alpha}.$$

This value is of the same order of magnitude as the intuitive value, ct_0, obtained for a flat space with no expansion if one calculated the distance light could travel from $t = 0$ until today ($t = t_0$).

We could equally have calculated the horizon at any time, t, in cosmic history. This would be necessary if we wished to determine whether or not two points in space were already causally connected. In this case the calculations above are still valid, but it is convenient to replace t_0 by t so that

$$\chi_H(t) = c \int_0^t \frac{dt}{R(t)}$$

and

$$d_H(t) = R(t)\chi_H(t).$$

It is important to realise that whichever model of the universe is chosen, the comoving value of the horizon is an increasing function of time, since the integrand of the integral above is positive (see figure 2.9).

3

The dynamics of the universe

3.1 The metric of the universe

3.1.1 The metric tensor

The metric is what mathematicians call a quadratic form. This means that it is a polynomial of degree 2 in the increments of the spatial and temporal coordinates such as dt, dx, dr, $d\chi$, $d\theta$ etc. Such an expression can involve many terms; squared terms like dx^2, dt^2, ... (as in the Minkowskian form), but also with coefficients that depend on the coordinates, and mixed terms such as $dtdx$ etc. Specifying the coefficients appearing in front of each of these terms defines the metric.

Mathematics provides what is called a tensorial formalism for writing down the metric. By numbering the coordinates ($t = x_0$ and x_1, x_2, and x_3 for the spatial coordinates), one can write the metric as a sum of quadratic terms of the type $dx_\mu dx_\nu$ where the indices μ and ν can take on the values 0, 1, 2, or 3 with coefficients denoted by $g_{\mu\nu}$.

The array of elements $g_{\mu\nu}$ is called a tensor, and behaves globally as one entity **g**. The tensor that interests us, the metric tensor, has 16 coefficients, although in fact only 10 are independent.

By means of Einstein's field equations, general relativity provides a means of (indirectly) calculating these unknown coefficients from the matter content of the universe and from an additional constant called the cosmological constant. These equations can be written in a condensed form as one equation between two tensors. However, since the metric tensor has 10 coefficients which have to be determined, the tensorial equation is equivalent to 10 scalar equations. Einstein's equation does not directly yield the metric tensor. It yields another tensor called the Ricci tensor, related to the metric tensor and involving the derivatives of the metric tensor with respect to the coordinates. The equation links this tensor to another tensor, $T_{\mu\nu}$, called the energy–momentum tensor, which characterises the contents of the universe (essentially determined by the density and pressure of matter and radiation).

General relativity and model universes

In order to describe the universe we must define its spatial geometry. Since we supposed this to be homogeneous and isotropic, we know that this geometry can be described by the form shown above. The only thing that remains to be done is to determine the value of k and the form of the function $R(t)$, which can be achieved by applying Einstein's equation.

The simplest case is where gravitation plays no role. It corresponds to a Minkowski universe, whose properties are essentially those of special relativity; the coefficients of the square terms reduce to c^2 or -1 and the others are zero.

Such simplicity is not normally found in general relativity, where the coefficients are obtained from Einstein's equations. These require knowledge of the energy–momentum tensor describing the matter content of the universe, and the cosmological constant, Λ. The contents comprise all forms of energy, whether this is in the form of ordinary matter or in the form of radiation, and other possible forms of matter not yet detected.

3.1.2   The cosmic fluid and its equation of state

In calculating the energy–momentum tensor for the universe, we shall invoke one further simplifying hypothesis. This is that the contents of the universe can be adequately described by what is called a perfect fluid, which means that the only relevant properties, from the point of view of the dynamics of the universe, are its density and pressure, which are the same at all points of the universe since it is supposed homogeneous. In this situation the energy–momentum tensor takes on a simple form and Einstein's equations reduce to a set of two differential equations. These provide the means to calculate both $R(t)$ and k as a function of the density, ρ, and pressure, p, of the 'cosmic fluid', as well as the cosmological constant Λ.

It is very natural that the density of matter in the universe should appear as a crucial parameter in determining the evolution of the universe. Effectively the universe evolves under the influence of its own weight: the higher its density the greater its gravitational self-attraction, and the more the expansion must slow down.

One can in general describe a perfect fluid by an equation of state* relating the pressure, p, to the density, ρ. In Einstein's equations the density term appears in the form ρc^2, where c is the speed of light, and pressure in the form p. It is thus relevant to compare the relative values of ρc^2 and p.

Cold matter

For ordinary matter pressure is given by $p \sim \rho v_{th}^2$ where v_{th} is the thermal velocity. Matter is defined to be 'cold' or 'non-relativistic' when v_{th} is small compared with c. In these conditions the pressure term is small compared with the density term, ρc^2, and can be ignored without making any serious error. In other words, for cold matter, it is convenient to make the approximation $p = 0$.

Matter can be considered to have been cold for the last few thousand million years, so that the approximation $p = 0$ is adequate for its description. Only in the earliest times, roughly the first million years, was this not the case.

Radiation

Electromagnetic radiation is present everywhere in the universe. Although its effect today is not very great, its dynamical role could a priori be important, and certainly was at the beginning of cosmic history. Its high pressure cannot be neglected. Its equation of state can be written

$$p = \frac{1}{3}\rho c^2.$$

These two forms of energy, matter and radiation, are the archetypical contents of the universe. The entire contents of the universe can be reduced to a combination of these two species.

The vacuum energy

Even so, a third possible form of energy has recently been introduced into cosmology. Recent results in quantum field theory have shown that this third form could have a dynamical effect in cosmology. It is a question of what one calls, somewhat unfortunately but graphically, vacuum energy*. By vacuum is meant a fundamental state relative to which one measures excitations of the field (see subsection 4.4.5).

Theory suggests that such a state has an energy density, ρ_{vacuum}, and behaves as though it has a pressure, albeit a negative pressure, obeying the equation $p = -\rho_{vacuum}$! We shall not discuss this equation any further, but one should remember it for later. We shall return to it when we discuss inflationary* models. It is remarkable that the contribution of this vacuum energy is formally identical to that of the cosmological constant if we put $\Lambda = 8\pi G \rho_{\text{vacuum}}$.

3.1.3 Friedmann's equations

If we use an energy–momentum tensor for a perfect fluid in Einstein's equations we obtain the fundamental equations of cosmology:

$$\frac{R''}{R} = -\frac{4\pi G}{3}\left(\rho + \frac{3p}{c^2}\right) + \frac{\Lambda}{3} \qquad (3.1)$$

and

$$\left(\frac{R'}{R}\right)^2 = \frac{8\pi G \rho}{3} + \frac{\Lambda}{3} - \frac{kc^2}{R^2} \qquad (3.2)$$

to which is sometimes added

$$\frac{d}{dt}(\rho c^2 R^3) = -p\frac{d}{dt}(R^3). \qquad (3.3)$$

This last equation can be deduced from the first two. These Friedmann equations are the very foundations of the big bang models, and determine the structure of the universe once its contents is given.

We should realise that even with these simplifying assumptions stated previously (homogeneity and perfect fluid), the density and pressure do not entirely determine the geometry and the dynamics of the universe. One also needs to know the cosmological constant. For the moment we shall leave aside this special parameter, and in what follows shall take it to be zero. Later we shall look at its possible effects in subsection 3.3.3.

3.1.4 The equation of state and density dilution

The third Friedmann equation (3.3) allows us to analyse the behaviour of the cosmological density as a function of time (or of the scale factor, which amounts to the same thing) provided we know the form of the equation of state. Intuitively, since volumes are subject to the expansion, and hence grow, matter within any volume is conserved in such a way that its density (that is the quantity of energy per unit volume) falls with cosmic time as the scale factor $R(t)$ grows. There is a dilution.

In the case of ordinary cold matter, which has dominated the universe for thousands of millions of years, the pressure is zero. In fact, whether we have a 'gas of stars or galaxies' or a gas of atoms and molecules, so long as matter is not relativistic the pressure is so small compared with the density that it can be ignored in Einstein's equations. Equation (3.3) is in this case easy to solve and gives $\rho R^3 = $ constant. In other words, the mass density of cold (ordinary) matter dilutes as R^{-3}.

This is easy to understand. Effectively if matter is neither destroyed nor created, the number of particles present must be conserved in an expanding (comoving) volume. The number of particles per unit volume, the number density, thus varies as R^{-3}.

The situation for radiation, formed by photons, is different. The number density of photons also varies as R^{-3}, but the energy of each photon is not constant because of the redshift of its wavelength, which varies as R^{-1}. Thus the energy dilutes as R^{-4} and not as R^{-3}. This result can of course be obtained from equation (3.3).

Finally, for the vacuum (with equation of state $p = -\rho c^2$), the density ρ does not dilute at all but remains constant in time as $R(t)$ varies.

The dynamics of the universe depends then on the form of the energy that dominates. We shall, in the first instance, suppose that it is just one form of energy that completely dominates so that we can examine the resulting dynamics. We shall then see that these cases do indeed constitute good approximations to the real phases of cosmic evolution.

3.1.5 The cosmological parameters

Normally one introduces Hubble's parameter or 'constant'

$$H_0 = \left(\frac{R'}{R}\right)_0,$$

which measures the present rate of expansion. It is related to a quantity we have already defined, the deceleration parameter

$$q_0 = -\left(\frac{R''R}{R'^2}\right)_0, \tag{3.4}$$

which measures slow down in the rate of expansion. (The suffix 0 always denotes the present value of quantities.)

3.2   The matter dominated universe

3.2.1   Matter in Friedmann's equations

Cold matter dilutes as R^{-3}. This allows us to write the density as

$$\rho = \rho_0 \left(\frac{R_0}{R} \right)^3 ,$$

where ρ_0 and R_0 are the values of the density and scale factor at the present epoch ($t = t_0$). One can then easily obtain the following:

$$q_0 H_0^2 = \frac{4\pi G \rho_0}{3} \tag{3.5}$$

and

$$\frac{kc^2}{R_0^2} = (2q_0 - 1)H_0^2 \tag{3.6}$$

so that Friedmann's equations finally give

$$R_0 H_0 dt = dR \left(1 - 2q_0 + 2q_0 \frac{R_0}{R} \right)^{1/2} .$$

Integrating this equation links any value of cosmic time with the corresponding value of the redshift z:

$$t = \frac{1}{H_0} \int_0^{1/(1+z)} F(x)\, dx$$

where we have used the relation

$$\frac{1}{1+z} = \frac{R}{R_0}$$

and defined the function $F(x)$ as

$$F(x) = (1 - 2q_0 + 2q_0 x)^{-1/2}. \tag{3.7}$$

We shall see in subsection 3.3.4 that this integral enables us to calculate the age of the universe by putting $R_{t=0} = 0$. We shall comment on this below.

Although the universe has not always been dominated by matter, without doubt this equation does describe many characteristics of the actual universe, and so merits a detailed study.

3.2.2 The initial instant

In the big bang models R goes to zero at some finite time in the past. Previously we chose the constant of integration in such a way that R was zero at t equals zero, although this was entirely for convenience. We could also have chosen year 1789, 1644 or 1776 as time zero.

This might be less misleading as one should remember that the validity of these equations does not reach back into the past right to the instant when R goes to zero. In fact, when the scale factor R becomes small, densities and temperatures become very high, and the validity of general relativity is not guaranteed. At very high densities general relativity should be replaced by another theory of gravitation, which is not yet at hand. Thus we know that our models are not valid for describing the evolution of the universe far into the past, but we have nothing viable that could replace them. So we should recognise the limits to our understanding, which is more and more speculative the further back in time one goes. Let us remember that in any case the evolution described by the simple models expounded here are only approximately valid, and then only down to values of R and t close to zero. Moreover, the instant zero itself is only a convenient extension of the model which certainly does not correspond to reality.

With this word of warning, we can now attempt to solve the equations in order to uncover the dynamics of the universe. Everything depends on the value of the quantity $1 - 2q_0$!

3.2.3 A flat universe

In the simplest case, for $q_0 = 0.5$, this quantity is zero, and from the equations above gives the present density of the universe as

$$\rho_0 = \frac{3H_0^2}{8\pi G}.$$

We shall come back to this point later. The equation above implies that $k = 0$, so that the geometry of this model is Euclidean, that is ordinary geometry. This model is called the Einstein–de Sitter model.

Solution of the equations determines the dynamics of the universe; the scale factor grows indefinitely as $t^{\frac{2}{3}}$. The present age of the universe (defined as the time that has passed since $t = R = 0$) is $\frac{2}{3}H_0^{-1}$. In order to avoid any epistemological confusion and not prejudge the possible existence of an earlier phase, whether similar or not to the present phase, we should remember this refers to the time that has passed since the present expansion phase began.

In these models the matter density has a value precisely equal to what is called the critical density, $\rho_{critical}$, given by

$$\rho_{critical} = \frac{3H_0^2}{8\pi G} = 1.96 \times 10^{-29} h^2 \mathrm{g\,cm}^{-3} \qquad (3.8)$$

where as defined previously, h is the value of Hubble's constant in units of $100\,\mathrm{km\,s^{-1}\,Mpc^{-1}}$. In fact, the matter dominated models divide into two classes according to whether the density is greater or smaller that this value.

3.2.4   Matter dominated models

If the density is less than the critical value, then the curvature parameter, $k = -1$, the universe is open, and space is hyperbolic. Solution of the dynamic equations shows that expansion takes place all the time, as in the Einstein–de Sitter model. This expansion slows down but never stops. According to observational data, this model is closest to the real universe.

If on the other hand the density is greater than the critical value, space has positive curvature and $k = +1$. The function $R(t)$ will not increase indefinitely, but reaches a maximum R_{max} at some time t_{max}. After t_{max} the function $R(t)$ decreases in a completely symmetrical way to its growth. This model gives a 'future contraction', contraction which lasts until the point at which the laws of general relativity once again break down, that is at the Planck era (see figure 3.1).

Eternal expansion or future contraction?

There are thus three classes of possible models dominated by matter. It is very likely that one of these models fits the present behaviour of the real universe. In order to decide which, it is useful to measure the present energy density of the universe, and compare it with the critical density, $\rho_{critical}$, defined above.

The present value of Hubble's constant is not known with any precision (and we have defined $h = H_0/(100\,\mathrm{km\,s^{-1}\,Mpc^{-1}})$). Taking into account this factor of 2 uncertainty in the value of H_0, the value of h for the actual universe lies in between 0.5 and 1, and the critical density is given by $\rho_{critical} = 1.96 \times 10^{-29}\,h^2 \mathrm{g\,cm}^{-3}$. The density parameter, Ω_0, is defined as

$$\Omega_0 = \frac{\rho_0}{\rho_{critical}} = 2q_0$$

for this class of matter dominated universes.

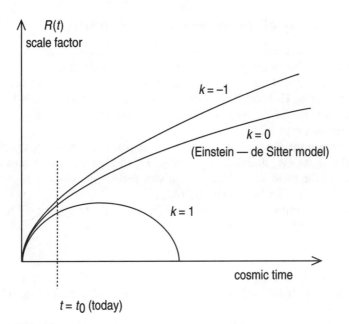

Figure 3.1: The scale factor as a function of time for the three classes of matter dominated universes.

We should also note that

$$\frac{kc^2}{R_0{}^2} = (2q_0 - 1)H_0{}^2,$$

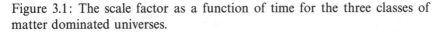

although this only valid if the cosmological constant is zero.

With the knowledge that the density of visible matter is in the region of 10^{-31} g cm^{-3}, less than the critical density, it is tempting to conclude that open models give a better description of the universe. However, it is possible that mass exists in an invisible form, so that the total density exceeds this latter value, and possibly attains the critical value. We shall discuss this possibility later.

These models are complete in the sense that they determine both the dynamics and the geometry of the universe. It should be noted that the three classes are distinguished both in their geometric and their dynamic properties; the models with spherical geometry are associated with future contraction, whilst those with no curvature or negative curvature are associated with continual expansion. It should be remembered that this association does not hold in models with non-zero cosmological constant. (Models with a cosmological constant are presented in subsection 3.3.3.)

3.2.5 Density of the universe and dark matter

Astronomers are familiar with the observable forms of matter, luminous stars (made of very hot gas) brought together in galaxies, gas, dust, planets etc. They can count the galaxies in the universe. The distance of each galaxy, estimated using Hubble's law, is only known up to the value of h. Hence the numerical value of the number density of galaxies (in galaxies per Mpc^3) is only known up to h^3.

Furthermore, for the nearest galaxies, astronomers have separately analysed the mass, M, and the luminosity, L, and established a mean value, $\langle M/L \rangle$, of the ratio M/L. Consequently to deduce the mass of a galaxy one can just multiply its luminosity by $\langle M/L \rangle$.

The true luminosity of a galaxy is found from its apparent luminosity, F, from a formula of the type

$$L = 4\pi \ F \ d^2,$$

where F is measured, but the distance d, which is obtained from the redshift, depends on h. It follows that we know the mass of each galaxy only up to a factor h^2.

Finally, the mass density of matter in the galaxies can be written

$$\rho_{\text{galaxies}} = N_{\text{galaxies}} \langle M/L \rangle \langle L \rangle$$

where the mean values are taken as typical for the population of galaxies in the universe. Thus we know ρ_{galaxies} to a factor of h^2. These estimates are subject to great uncertainty, but it would appear that ρ_{galaxies} lies in the neighbourhood of $5 \times 10^{-31} \ h^2 \mathrm{g \, cm}^{-3}$, which is only a few hundredths of the critical density, ρ_{critical}, and so corresponds to a value of Ω_0 around 0.01. Thus (if one believes that the cosmological constant is zero), the universe would have negative curvature and infinite spatial extension, and is destined to expand for ever.

Nevertheless, the analysis of the dynamics of galaxies and clusters of galaxies and the models of their formation indicate that there is more mass than we see; besides the visible matter there is dark matter. For each galaxy or cluster of galaxies, the total mass would appear to be about 10 times the visible. As a result, the total mass residing in these objects could be as high as 0.2 of the critical value. One should note also that this value is not above the limit imposed by models of primordial nucleosynthesis (see section 4.3.4) for the maximum possible value for the mass density of ordinary baryonic matter. Thus the most likely vale of Ω_0, for baryons, is in the region of 0.2. We should note that the existence of this dark matter does not change (if $\Lambda = 0$) the conclusion that the universe is open, its volume infinite, and its curvature negative, and that expansion will continue for ever.

Non-baryonic dark matter?

Some people refuse to believe in such a model universe, and would like a flat universe. They would therefore like Ω_0 to be equal to 1. To this end they take the amount of dark matter to be still greater, and to be in some unknown form, invisible and existing outside galaxies and clusters of galaxies. Although such a hypothesis seems unlikely, there is no reason for dismissing it a priori.

However the models of primordial nucleosynthesis are incompatible with a contribution from baryonic matter making Ω_0 greater than 0.1 or 0.2. A higher value would imply that the universe is dominated by a still unknown species of particles. Astrophysicists believed for some time that neutrinos could play this role, but the recently established upper limits on the mass of these particles have dashed any hopes they might have had. Apart from the fact that the nature of this species of particle is still a mystery, too high a value of Ω_0 is difficult to reconcile with a number of observations (most notably as we shall see later, the age of the universe). Despite this, the existence of new particles is not incompatible with some current ideas in particle physics. Moreover, it seems as though it would be easier to understand how galaxies formed if such particles existed (see chapter 5). The debate is not closed.

It is still the case that the 'desire' for a flat universe necessitates the introduction of supplementary hypotheses to the standard model (more complicated nucleosynthesis, the existence of exotic particles ...). This illustrates to what point cosmology, even scientific cosmology, rests on principles that lie outside the realm of science and more in the realm of metaphysics.

The search for dark matter

Finally, in some sense, there are two problems with dark matter. On the one hand there is a body of observations that suggests that in galaxies and clusters there is about ten times the mass that is seen. All this mass would give a value of Ω_0 of about 0.1. Furthermore there are a number of more speculative arguments that lead to a still higher value of Ω_0, and that there might exist a second component to this dark matter. In this case, and in order to be in agreement with analyses of the dynamics of these galaxies and clusters, this component must be distributed outside these systems, possibly around them, but in far distant regions.

At present we do not know the nature of this dark matter. As far as the component distributed in galaxies and clusters is concerned, there are very few constraints, so long as it is in agreement with models for primordial nucleosynthesis. This component could be either baryonic or non-baryonic. Nevertheless, it is very likely that a part of it at least is baryonic. The simplest hypothesis is that it is all baryonic (although possibly existing in several forms). As for the additional component ($\Omega_0 = 1$), if it really exists, it could not be baryonic, unless one assumes that primordial nucleosynthesis did not take place in accord with the standard scenario.

Astrophysicists and particle physicists try to imagine what form the dark matter could take, and to find this out. With respect to non-baryonic matter, current ideas in particle physics suggest several possibilities such as massive neutrinos, supersymmetric counterparts of ordinary particles (gravitinos, neutralinos, photinos), axions, WIMPS (Weakly Interacting Massive Particles), etc. Various groups have constructed or are in the process of constructing detectors designed to detect some of these particles and to see whether they exist and whether they could constitute a significant part of dark matter. So far these attempts have remained unsuccessful, although they have eliminated some candidates. These experiments continue today. The current development of bolometers to be used as detectors seems particularly promising.

On the other hand, with regard to the baryonic component, which almost certainly exists, astronomers have concluded that the most likely possibility is that this is in the form of objects intermediate between stars and planets. This is not the only possibility. Large amounts of very cold gas, and very massive objects (black holes) are sometimes proposed. However, taking into account both observational and theoretical constraints, this is the most likely. These condensed objects could, for example, have been formed from the contraction of clouds of gas whose mass was insufficient to form stars in the proper sense of the word; this condensation would not have provided sufficient temperatures to set off the nuclear reactions that provide the energy for stars. They would have become 'almost stars', or, as they have been dubbed, 'brown dwarfs'. Each one could have a mass of a few hundredths of the Sun's mass, and they could be sufficiently numerous to provide, if not the entire dark matter in the galaxies, an important part of it.

If this is the case, then the halo of our own Galaxy (the Milky Way) must be filled with compact objects, which are sometimes called MACHOs (MAssive Compact Halo Objects). If our halo is filled in this way with MA-CHOs, they must exercise some gravitational effect on light rays traversing the halo. Moreover, from time to time it must happen that light rays emitted by a distant star pass close to a MACHO before reaching us. In this case, the gravitational effect of the compact object (according to the predictions of general relativity) would be to bend the light rays, and moreover to instantaneously amplify the brightness of the star. For example, if the halo is filled with MACHOs, there is about one chance in a million that the brightness of a star in the galaxy closest to ours, the Large Magellanic Cloud, be observably amplified in this way. This is not a very great probability, but if one patiently observes several million stars, one should finally succeed in observing this effect.

A number of physicists considered that such a project could be successful. Two groups were set up, EROS (Experience de Recherche d'Objets Sombres, i.e. Experiment to Find Dark Objects) in France and the other, MACHO, principally in the USA, to undertake a systematic study of millions of stars in the Large Magellanic Cloud. This involved measuring and recording every night, using electronic detectors (CCDs or Charge Coupled Devices), the luminosities of millions of stars to see whether they displayed the predicted effect. The difficulty of this enterprise lay in the enormous amount of information that needed to be recorded and analysed, a double challenge which these teams were able to overcome. Thus after three years of taking data, the search proved fruitful. The EROS team detected two candidates, and the MACHO team a third. Thus after millions of stars had been examined, the light from three was seen to grow and then fall back to its previous brightness. This phenomenon lasted for about thirty days. The variation in the stars' brightnesses agreed with the predictions of gravitational lensing, and was shown to be the same for both blue and red light. This corresponded exactly to the predictions.

In the absence of any other possible explanation, it seems that brown dwarfs have been detected for the first time. If we take into account the small probability of this phenomenon, we are led to the conclusion that our halo is filled with these objects, and that they constitute a large part of the dark matter. All this remains to be confirmed by recording further data. On the other hand, the search for other candidates, whether or not baryonic, continues. In any event, it seems as though this question, which has tormented astronomers since the 1930s, will be resolved before the end of the millenium!

3.3 Big bang models

3.3.1 Dynamical equations

Matter dominated models describe the geometry and the dynamics of the universe in the present epoch and during the last few thousand million years. We should also study radiation dominated models, which describe the primordial, or very early, universe. The most general homogeneous and isotropic models should likewise take into account the possibility of a cosmological constant.

It is convenient to distinguish the various contributions to the pressure and density in Friedmann's equations according to the equations of state and laws governing the dilution. We defined for matter the parameter

$$\Omega_0 = \frac{\rho_{0,\text{matter}}}{\rho_{\text{critical}}}$$

Define a similar parameter for radiation:

$$\Phi_0 = \frac{\rho_{0,\text{radiation}}}{\rho_{\text{critical}}}. \tag{3.9}$$

and a 'reduced' cosmological constant

$$\lambda = \frac{\Lambda}{8\pi G \rho_{\text{critical}}} = \frac{\Lambda}{3H_0^2}. \tag{3.10}$$

Finally we introduce the notation

$$x = \frac{R}{R_0} = \frac{1}{1+z}.$$

From this the general relation follows

$$2q_0 = \Omega_0 + 2\Phi_0 - 2\lambda,$$

$$\Omega_0 + \Phi_0 + \lambda - 1 = \frac{kc^2}{H_0^2 R_0^2}. \tag{3.11}$$

Friedmann's equation can then be written as

$$\frac{x'^2}{H_0^2} = F^{-2}(x) = \frac{\Omega_0}{x} + \frac{\Phi_0}{x^2} + x^2\lambda + 1 - \Omega_0 + \Phi_0 + \lambda, \tag{3.12}$$

where all the quantities except x are constant in time. Integration of this equation provides the dynamics of the universe.

The variable $x = R/R_0$ grows with cosmic time. The equation above shows us that in the primordial universe (x small) radiation dominates and matter and the cosmological constant can be neglected, as can the curvature term.

3.3.2 The radiation epoch

In the first part of its evolution the universe's dynamics are dominated by matter, if the cosmological constant is zero. The epoch of matter starts at around one million years, and therefore has endured for about fifteen thousand million years. But during the first million years, dynamically speaking, it was dominated by electromagnetic radiation. So long as radiation dominates, the function F is approximately equal to $x\Phi_0^{-1/2}$ so that t varies as R^2 (R grows as $t^{1/2}$). Density dilutes as R^{-4} and temperature falls as R^{-1}.

3.3.3 The universe with a cosmological constant

Equations (3.11) also describe models in which the cosmological constant is not zero. Comparing the different terms one can see that the cosmological constant will only affect the dynamics when x is not too small, that is to say during the epoch of matter domination. In such a model the dynamics are therefore not modified during the first instants, and the results of the previous paragraph are still valid. Only the matter dominated epoch is modified and we obtain

$$2q_0 = \Omega_0 - 2\lambda \qquad (3.13)$$

and

$$\Omega_0 + \lambda - 1 = \frac{kc^2}{H_0^2 R_0^2} \qquad (3.14)$$

in place of equation (3.6).

In this case the dynamics will depend on the relative importance of the matter terms to the cosmological constant. Models with a cosmological constant are important for several reasons. Evidently, the first is that it is possible for the cosmological constant not to be zero, and therefore influence the present and recent dynamics. However there are also historical reasons; models without matter but with a non-zero cosmological constant were studied at the beginning of the century and went under the name of de Sitter models. Their dynamics are obtained by putting the matter term in the equations equal to zero. This gives an exponential, and not a power law, growth of $R(t)$.

Besides this, as we have already remarked, the effect of a cosmological constant can be identified with that of a vacuum energy. It is possible that during the first few instants of the big bang the universe was dominated by such a form of energy. Its dynamics would then have been identical to those of a universe dominated by the cosmological constant. This is the so-called inflationary model, which we shall describe in subsection 4.4.5.

As opposed to the density and pressure of matter which enter into Friedmann's equations, the contribution of the cosmological constant to the dynamics of the universe makes no reference to its contents. This unsatisfactory situation goes against the spirit of one of the most positive aspects of general relativity, according to which it is the material content alone that determines the dynamics. For this reason it is reasonable to think that this constant should not be included, or in other words, should be taken to be zero. This for example was the view of Einstein who introduced the constant into his equations on noting that this was the only way of achieving stationary models of the universe. Today we know from observation that the universe is not stationary and this argument no longer stands. Supporters of a zero cosmological constant nevertheless could be embarrassed by another argument connected to the age of the universe.

3.3.4   Age without age

Friedmann's equations allow one to estimate the scale factor $R(t)$ as a function of time so long as one knows the relative proportions of radiation and 'cold matter', and of the cosmological constant. Knowing this constant one can calculate the 'age of the universe', defined by the time that has elapsed since the scale factor was zero, i.e. $R = 0$.

First of all, in some cosmological models where the cosmological constant is non-zero $R(t)$ is never zero in the finite past. In this case the universe will be said to have an infinite age, and the models will be said to be 'without a big bang' (see figure 3.2).

However here we shall only be interested in big bang models, which we believe describe the universe better, and which are defined as those in which $R(t)$ goes to zero in the finite past. Once again, to avoid any confusion, let us remember that this disappearance does not correspond to a reality, but only to a mathematical property whose importance should not be exaggerated.

Equation (3.11) describes the evolution of the universe as a function of cosmic time, and assumes its contents to be in the form of identifiable forms of matter, radiation as well as a cosmological constant (and ultimately a vacuum energy). In these big bang models general relativity predicts smaller and smaller values of the scale factor, $R(t)$, the more one goes back in time. The definition of the age of the universe given above implies that eventually the scale factor goes to zero. However, when the scale factor is very small so will be all cosmological dimensions, and cosmological matter will be so dense that the known physical laws cannot be applied in these instances. In particular, general relativity cannot be used to describe the dynamics of the universe and the big bang models lose their validity. Thus, as we have already said, it is not permissible to extrapolate the expansion back beyond this moment. To extrapolate to $t = R = 0$, implying infinite densities, has no physical meaning.

Nevertheless, just for expediency, we choose to date events as though the law of expansion can be extended to this instant. In other words, we choose time zero to be, by definition, the moment when the scale factor goes to zero as we trace the expansion (fictitiously) back into the past. In this conventional time scale, the density attains the Planck value at a time called the Planck time, t_{Planck}, which has the value 10^{-43} s (see figure 3.3).

Hence the timescale we have defined only has a meaning after t_{Planck}. Any application of the cosmological models before the Planck time is therefore

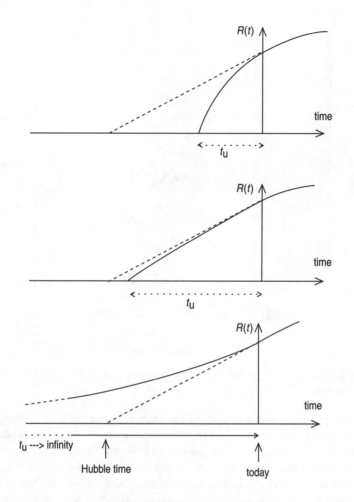

Figure 3.2: The age, t_u, of the universe for the three cosmological models. In the last case (in which the cosmological constant is non-zero) the age is infinite.

absolutely false. In particular, it is of course *an abuse of language to talk of the big bang as the beginning, or creation of the universe.*

What meaning then can we give to this definition of the age of the universe? It is simply the time that has passed since the beginning of the expansion phase which concerns us. The Planck time is so small compared with the thousands of millions of years of the universe's evolution that one would make only a tiny error in ignoring it! If there were another earlier

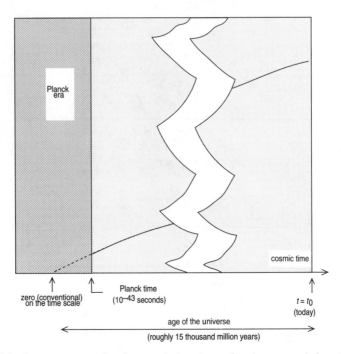

Figure 3.3: Once zero on the time scale has been fixed we can define the age of the universe.

phase of expansion in the history of the universe, or a phase of a completely different nature, its duration is not taken into account. The term 'age of the universe' is then a misuse of language and it would be wrong to associate it with the time since some 'birth' or 'creation' of the universe. Above all the definition is an operational one, because according to the big bang models, every known object, whether astronomical such as stars, galaxies etc., or microscopic objects such as atoms and nuclei, molecules, or crystals, only came into existence long after the beginning of this phase.

Knowing what might have happened before t_{Planck} is a different problem. Given the time scale that we have chosen, we would have to give these events negative time coordinates (which would be no more mysterious than dating events before the birth of Christ). This simply underlines the fact that the time scale we have chosen is a matter of convention.

Mindful of these cautionary remarks, how can we estimate the age of the universe, t_u? According to the definition, t_u is given by the equation

$$t_u = \frac{1}{H_0} \int_0^1 F(x)dx, \tag{3.15}$$

where $F(x)$ is defined in equations (3.7) and (3.12) and depends on the constants Ω_0, Φ_0 and λ_0.

3.3.5 What is the age of the universe?

We know that the radiation parameter, Φ_0, is at most one thousandth of Ω_0. As a result its contribution to the integral is negligible. One can also express this in the following terms: the universe was dynamically radiation dominated for less than 10 million years but was dominated more than 10 thousand million years by matter. It is thus legitimate to neglect the duration of the radiation epoch in the calculation of the age of the universe. The small error committed in neglecting this term consists in replacing for a very small fraction of the time (one ten thousandth) the expansion law $R(t) \sim t^{1/2}$ by $R(t) \sim t^{2/3}$.

In the case where the cosmological constant is zero one easily finds

$$t_{\mathrm{u}} = \frac{1}{H_0} \int_0^1 \left(\frac{\Omega_0}{x} + 1 - \Omega_0 \right)^{-\frac{1}{2}} dx.$$

It is easy to confirm that the age obtained is necessarily less than the Hubble time H_0^{-1}, say $9.78 \times 10^9 h^{-1}$ years. For $\Omega = 1$ it is exactly $6.52 \times 10^9 h^{-1}$ years. If Ω is greater than 1 the age is less, say $7 \times 10^9 h^{-1}$ years. If Ω is less than 1, the age is greater, although less than $9.78 \times 10^9 \ h^{-1}$ years.

Evidently, if the cosmological constant is not zero, these values are changed, in one direction or another according to its sign (about which we know nothing a priori). It is then necessary to consider the more general formula given above.

The astronomical age

The expansion of the universe is the most compelling evidence that the universe evolves. It is therefore not surprising that the age of the universe should be defined through this expansion, and furthermore linked to the fundamental cosmological parameters (Hubble's constant, matter density of the universe, and the cosmological constant). Yet these are still poorly determined. Another type of measurement of the age of the universe could put constraints on their values. Although there is no such measure, the expansion phase must be at least as old as the objects in the universe. Any lower limit on the age of such an object must also be a limit on the age of the universe.

The first limit derives from the age of the Earth, which is about 4.4 thousand million years. Compared with the numbers mentioned earlier, this is not very restrictive.

With the aim of establishing more serious limits, astronomers attempt to find the age of the oldest stars. Their ages appear to be comparable if not greater than the Hubble time (according to the value of Hubble's constant adopted). These measurements could then indicate that the cosmological constant is not zero, or that the density of the universe is less than the critical density.

Two methods can be used to estimate the ages of stars. One is based on models of stellar evolution, and the ages predicted by these models. The other depends on dating the stars from radioactive isotopes, in much the same way as carbon 14 dating works. Astronomers are not in agreement over the ages of the oldest stars, although they are reckoned to be in between 12 and 20 thousand million years old.

It is helpful to recall the following facts. For a fixed value of the cosmological constant, the age of the universe decreases both with Ω_0 and H_0. Let us recall a few values for a zero cosmological constant. $9.78 \times 10^9 h^{-1}$ years is an upper limit, no matter the value of Ω_0. If $\Omega_0 = 1$ the upper limit is $6.52 \times 10^9 h^{-1}$ years. It is interesting to compare the ages of the stars with these limits in order to obtain upper limits on Ω_0 and H_0.

For example, an age greater than 13 thousand million years implies either a value of Ω_0 less than 1, or a non-zero cosmological constant, no matter what value H_0 takes, provided it is greater than $50 \, \mathrm{km \, s^{-1} \, Mpc^{-1}}$. Yet astronomers often talk of stars as being as old as 15 thousand million years! No doubt they are still not sufficiently sure of their estimates to give us really definite limits, but they are not so far away from excluding models with zero cosmological constant, or with density equal to the critical density.

4

The primordial universe

4.1 Towards the big bang

4.1.1 Expansion, dilution and cooling

Up to now we have given a relatively geometric (and dynamic) description of the universe, without paying too much attention to the material contents of the universe. The essential upshot has been that every cosmic length increases with time in proportion to the scale factor $R(t)$, and volumes in proportion to $R^3(t)$, and consequently there is a dilution of all forms of matter and radiation in the universe.

A given quantity of matter is distributed within a given volume which grows as $R^3(t)$. Hence the number density of particles (the number of particles per unit volume) decreases as $R^{-3}(t)$. Recall that the energy density of non-relativistic matter decreases as $R^{-3}(t)$ (and that of relativistic matter or radiation as $R^{-4}(t)$). As a result of expansion the contents of the universe must undergo an evolution, and consequently the universe has a history. Are we able to reconstruct this history? It is precisely this that model cosmologies or big bang models we describe here do.

The further we go back into time (and thus approach the Planck time t_{Planck}) the denser the universe becomes. By the same token, the hotter the universe becomes. These are the first traits of the universe according to the big bang models.

4.1.2 The absence of structure

Far into the past (during the first thousand million years roughly), neither galaxies nor stars existed. Further still (during the first million years or so), the temperature was so high that even atoms could not exist without immediately being broken up into their constituents, nuclei and electrons. Yet still further (during the first few minutes) the nuclei themselves could not withstand the high temperatures and densities of the primordial universe: it was only later in the current cosmic history that they appeared. And there is no doubt that certain elementary particles have not always existed as they now are. Our physical understanding does not allow us to reconstruct the whole of history up to the Planck time. There certainly has been an expansion, dilution and cooling since that time. But we still do not have the necessary tools to know exactly the properties of the universe at these times in the distant past. Physicists know that their model is not valid for times too far back into the past.

However, the reconstruction of the history of the universe does highlight the great importance of electromagnetic radiation in the past. There are two reasons for this. On the one hand, the way in which energy density diminishes is different depending on whether one is talking of matter or radiation. The energy density of electromagnetic radiation dilutes more quickly (as $R^{-4}(t)$) than that of matter (as $R^{-3}(t)$). As a result, the further one goes back in time, the more important is the energy density of radiation compared with matter. Throughout the primordial universe (about the first million years) the density of radiation essentially determined the dynamics of the universe (see figure 4.1).

On the other hand, during roughly the same period, radiation interacted strongly with matter. As a result the primordial universe was opaque, and stars and galaxies could not condense, as we shall see later.

It is only after this first million years, after the end of the primordial era (at a temperature of about 4000 K) that matter in the universe became transparent. From this moment on, local contraction (compared with global expansion) set in, and resulted some hundreds of millions of years later in the appearance of the first stars and galaxies.

4.1.3   Thermal equilibrium

During the first primordial period (corresponding to the first million years or so), the universe was hot and dense. Matter emitted electromagnetic radiation in the same way as an incandescent body emits visible light. However, the matter in the universe, which was hotter than any imaginable incandescent object, emitted a much more energetic form of radiation than visible light. More accurately, we should say that matter was in equilibrium with radiation, with photons continually being absorbed and emitted by this gaseous matter. Matter was thus bathed in radiation, with which it constantly interacted. As a result, for example, matter and radiation maintained

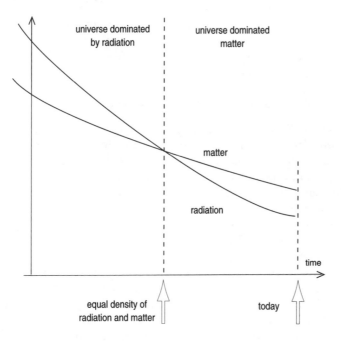

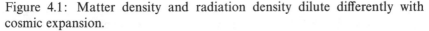

Figure 4.1: Matter density and radiation density dilute differently with cosmic expansion.

the same temperature. (The temperature of radiation is characterised by its energy or spectral distribution. Thus an incandescent body emits red and then white (yellow) light when it is heated to higher temperatures. At lower temperatures it emits infrared. At higher temperatures it emits ultraviolet light, X-rays, and ultimately γ-rays.)

It is an essential property of big bang models that they state that the contents of the universe were hot and dense at the beginning of its history. It is in precisely these conditions that the reaction rates for every sort of reaction are very high. These incessant reactions made it possible, during the greater part of the primordial era, for the various species of matter and radiation to interact and so remain in mutual equilibrium, thermal equilibrium. If one goes back in time far enough (for example to a temperature of 10^{12} K, at time $t = 10^{-5}$ s, say) every species of particle was in equilibrium. In this case, their abundances and all their properties can be calculated from laws of quantum statistics. At this time the state of the universe can be well understood since it is relatively simple thanks to this equilibrium. This instant provides a convenient starting point for calculating the events that follow.

The history of the universe is characterised by a progressive breaking of this equilibrium. Little by little, in proportion to the fall in the density and temperature, different species of particles, one after the other, moved away from thermal equilibrium.

4.1.4   The radiation dominated universe

The function $R(t)$ is of fundamental importance for following the evolution of the universe. Effectively cosmic distances increase as R and volumes as R^3; the density of matter decreases as R^{-3} and of radiation as R^{-4}. The temperature of radiation in the primordial universe decreases as R^{-1} (for thermal radiation, density and temperature are linked by the laws of statistical mechanics). Friedmann's equations allow one to calculate this function $R(t)$ describing the dynamics, the laws of dilution etc., as we discussed before. The case of the primordial universe corresponds to radiation dominated dynamics.

The game, then, consists of calculating which species of particles are in equilibrium at any given time, and which have left equilibrium. At 10^{-5} s, as we have already said, every species of particle is in equilibrium. The laws of statistical mechanics tell us all their properties (except perhaps for various fluctuations which we shall look at in a second order analysis). Starting from this instant of known equilibrium, we can reconstruct the consequent unfolding of universal history, the evolution of the universe.

What different species are present? The answer to this important question does not follow immediately. Clearly we have to consider the components of ordinary matter: protons, neutrons, and electrons. To these we have to add their antiparticles, antiprotons, antineutrons, and positrons. These do not play (in all likelihood) any cosmological role today, but this was not necessarily always the case. We also have to add to these particles we know to exist without knowing necessarily a priori whether they played an important cosmological role; this must at least be examined. This is the case for the three species of neutrino (electron, muon, and tau neutrinos) and their associated antiparticles, muons and pions etc. Of course we have to consider photons, the particles of electromagnetic radiation. Finally it is possible that there are other still unidentified particles (let us call them provisionally *cosmions*) that exist or existed and that play an important cosmological role. It is useful in this case to try to imagine what their properties might have been.

4.1.5   Coupling matter to radiation

The matter that we observe today is made up of baryons (protons p and neutrons n) possibly occurring in the nuclei of atoms, but not necessarily, and electrons. These different species, which are coupled by electromagnetic and nuclear interactions, behave in the first analysis globally, as one sole component which we shall call *baryonic matter*. It is essential to look at its coupling with radiation.

From the fact that thermal equilibrium existed early in the universe, both baryonic matter and radiation can be described precisely by thermal distributions, at a common temperature, T_u. So long as this equilibrium lasted we can talk of T_u as the temperature of the universe. However this concept loses its meaning when the equilibrium is broken.

The thermal equilibrium between matter and radiation is assured by the continual interaction between photons and the material particles. In today's conditions, electrons are bound to nuclei to form electrically neutral atoms (which in turn combine to form molecules, crystals and rocks etc.). But at the very high temperatures of the primordial universe, matter was ionised: the electrons were separated from the nuclei.

In the primordial universe it was the ionisation that assured the interaction between photons and matter, which in turn produced the equilibrium between radiation and matter. Effectively, photons of the radiation field collide with free electrons of the ionised gas. They interact (scatter) and so cannot propagate freely. The ionised gas does not allow the radiation to pass; it is not transparent but opaque. Furthermore, the interaction between photons and electrons assures a total coupling between radiation and matter.

So long as matter in the universe was ionised, it was coupled to radiation. Since thermal equilibrium was thus guaranteed, the two components had the same temperature and the same dynamical evolution. In particular, matter could not condense without also taking with it the radiation. The latter, however, because of its very high pressure, had an enormous capacity to resist compression. Its presence prevented the early formation of galaxies, as we shall see.

4.1.6 Decoupling and recombination

Matter remained ionised during roughly the first million years of cosmic history. The point at which ionisation ceased marks the end of the primordial universe. Electrons which were formerly free recombined with protons to form neutral atoms and the abundance of free electrons drastically decreased. Yet they were indispensible for equilibrium, assuring the coupling between matter and radiation. Their disappearance marks the end of this coupling because the photons (long wavelength) interact very feebly, almost not at all, with neutral atoms.

After the recombination* of electrons, photons were able to freely propagate without interaction through the neutral cosmic gas, which was now transparent. Moreover, the coupling between matter and radiation ceased, and each component followed from this point its own separate history. This was a truly dynamic decoupling of great import. It separates the history of the universe into two phases. During the first, the primordial universe, matter was ionised, opaque to electromagnetic radiation, and in thermal equilibrium with this radiation. This matter was essentially composed of protons, neutrons, and electrons (and after the first few minutes, nuclei of helium, that is to say a structure of protons and neutrons bound by nuclear forces). These particles and the radiation were at a common temperature, T_u, which fell as R^{-1} owing to expansion. The second phase is the post-recombination universe.

When the temperature T_u had fallen to just above a certain limit, T_{rec}, ionisation was no longer guaranteed, and the electrons and protons recombined. This recombination marks the end of the primordial universe. Its occurrence is determined by the moment when T_u fell to T_{rec}, about 4000 K (K stands for kelvin*, or degrees absolute). In fact the laws determining the degree of ionisation in thermal equilibrium predict that more than 99% of hydrogen will be ionised at 5000 K whereas at 3000 K only less than 1% is. In what follows we shall adopt the value of 4000 K, whilst remembering that recombination and decoupling is not an instantaneous or complete process.

The most convenient way to describe the time of recombination – the separation between the primordial and the recent universe defined by the value of T_{rec} of temperature in cosmic history – is in terms of the corresponding value of the redshift, z_{rec}, which is approximately 1500. The precise time depends on the particular cosmological model chosen, but is in the region of one million years.

It is then at about $z = 1500$ that matter in the universe becomes neutral. Henceforth, decoupled, matter and radiation followed separate histories. In particular, matter, which up to now had been prevented from condensing by the radiation pressure to which it had been linked, was able to begin to contract in a complex process that ended, hundreds of millions of years later, in the appearance of the first stars and galaxies. Later on we shall study the various stages in this process, but first of all let us discuss what happens to the radiation. It must have evolved in such a way that it remained observable in the form of the microwave background, which was detected and found to have precisely the expected properties. This is one of the most resounding successes of the big bang models (see figure 4.2).

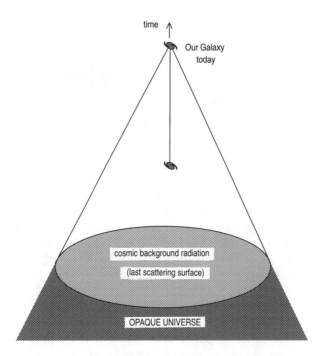

Figure 4.2: The cosmic background radiation is emitted within a sphere (a disc in our schematic figure) in the past, and called the sphere of last scattering.

4.1.7 The cosmic background radiation

How did radiation evolve after recombination? Since from this moment on it hardly interacted at all with matter it propagated freely, the universe appearing to it as a vacuum. So nothing happened to it, except that it experienced the cosmic expansion. It had no, or very little, structure, and continued to have none.

After recombination, and in the absence of any interactions, there was neither disappearance (absorption) nor creation (emission) of photons. Their density, the number of photons per unit volume, varied as R^{-3}. This can also be expressed as their comoving number density being constant. However, the radiation does experience a redshift, and so the frequency (and therefore the individual energy) of each photon decreases as R^{-1}. The distribution keeps the same form but is shifted as a whole to longer wavelengths; it maintains extremely closely a black body distribution but with a temperature given by

$$T(t) = T_{rec} \frac{1 + z}{1 + z_{rec}}.$$

A rigorous calculation shows fairly easily that radiation, when left to itself without undergoing any interaction with matter, keeps its thermal spectrum even though it is not in thermal equilibrium with matter at a temperature that decreases as R^{-1}. Although it cools, it continues to fill the universe.

The observed radiation

At the time of recombination the universe was bathed in radiation at about 4000 K, and therefore essentially composed of photons corresponding to visible and infrared light with a typical energy of the order of one electron volt* and wavelength of the order of 0.1 microns (one micron (μm) is 10^{-6} m). The calculation above implies that today the energy is much less, and that the temperature is given by

$$\frac{4000}{1 + z_{rec}} K.$$

If z_{rec} is of the order of 1500, these photons must be in the radio waveband with a wavelength of around one millimetre.

Radio astronomers do observe a background radiation which is extremely homogeneous and isotropic, and which has a thermal (black body) spectrum, with precisely the properties predicted by the big bang model. This brilliantly confirms the big bang. The radio astronomers Arno Penzias and Robert Wilson received the Nobel Prize for their discovery. The temperature of this Cosmic Microwave Background radiation or CMB, which we shall use for short in future, has been measured extremely accurately. The most recent measurement has been carried out with the American satellite COBE (COsmic Background Explorer) and gives the value of 2.726 K, with a very small error. This radiation, according to the standard interpretation, has its origin at a redshift $z_{rec} = 4000/2.726$ which is about 1500. It represents the oldest trace that astronomers have recorded. Recall that it propagated totally freely since the time of decoupling until today. The properties that we discover it has today (deviations from purely thermal radiation, anisotropies) should give us a complete fossil record of this distant instant. This is why astronomers observe it with so much determination now. Having in the first analysis confirmed the big bang, it should reveal to us more information about its nature (see chapter 5).

We should note, in passing, that all forms of matter and the structures that we know were born well after recombination, that is the last moment at

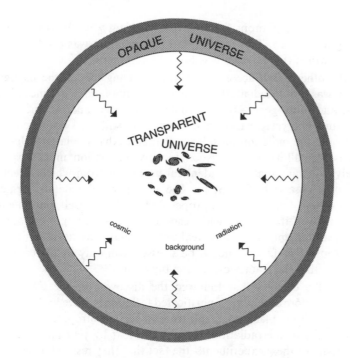

Figure 4.3: The cosmic background radiation viewed spatially.

which this radiation interacted with matter. These structures, then, have not affected the radiation. Nevertheless, if the radiation has not interacted with matter since $z = 1500$ it is because matter ceased to be ionised, as we have already stated. It is however not completely impossible that matter might have been re-ionised locally. Such is the case for example in the interiors of clusters of galaxies. Perhaps it was also the case in other circumstances. Thus it is possible that the CMB carries an imprint of such post-recombination re-ionisations. This of course complicates the analysis of its properties (see figure 4.3).

In the years that followed the discovery of the background radiation, many observations revealed two principal features: its thermal character and its isotropy. Both corresponded precisely to the predictions of the big bang, whilst no other consistent explanation could be found. Over and above the mere existence of the CMB, these results strongly vindicated the big bang model. The thermal character of the radiation – confirmed by successive observations at different wavelengths in the mm range – was borne out by the form of the spectrum, which agreed with a blackbody radiation law:

the energies of the photons forming the spectrum were distributed exactly according to this law. This implies that the radiation was produced during a phase in which the universe was hot and very dense.

On the other hand, according to the big bang model, the background radiation was produced at the time of recombination, a time when the galaxies, and more generally all structures, whether microscopic or astronomical, did not yet exist. The universe was still very homogeneous, as indeed would the radiation have been. It follows that the latter, which must have conserved all its characteristics since recombination, must now appear extremely isotropic, that is, be identical no matter from which direction in the sky it is observed. Except for one very particular detail, termed the *dipole anisotropy* (which is not a property of the CMB itself, but an effect of the movement of the Galaxy with respect to the universe), more and more precise measurements failed to reveal until recently any deviation from this high isotropy. This fact, it should be stressed, can only be understood and explained within the framework of the big bang.

During the decades that followed the discovery of the CMB various groups, using more and more sophisticated techniques, have looked either for deviations from the black-body spectrum or for anisotropies or fluctuations in the intensity of the radiation. Despite the year by year improvements in the sensitivity of these experiments, the fact that they resisted detection gave greater and greater confidence in the big bang model.

Deviations from a thermal spectrum

It has sometimes been suggested that it is going too far to consider these results as confirmation of the big bang. On the other hand, considering that no other model is capable of explaining these characteristics, this view point is quite justified. Furthermore there is one episode in the history of the observations of the CMB that should convince one of this. At the beginning of the 1980s a group of American and Japanese observers announced that they had detected an excess of radiation compared with the thermal spectrum (in the sub-millimetre wavelength range) implying a deviation from the predictions of the big bang. There was great excitement in the astronomical community. Was it an unimportant observation, explicable within the framework of the big bang by simply making a few additional assumptions, but not altering its fundamental premises? Calculations showed that one could not be so optimistic. These results, which appeared to be really incompatible with the big bang models, posed a genuine problem. Later it turned out (and this was confirmed by the COBE satellite) that the observations were wrong. But the anecdote does demonstrate to what point the confidence that astrophysicists have in the big bang is based on observations: just one result that does not fit can completely shake this confidence. On the other hand, each result in agreement is stronger confirmation.

Of course the big bang (and more generally any cosmological model) can only account for the global properties of the universe. It is necessary to provide a more detailed description to account for galaxies, stars and all observed cosmological objects and systems. However, the presence of these objects and the constraints that have to be imposed for their formation slightly modify the expected properties of the CMB. Even if the overall structure of the continuous spectrum should be black body or thermal, small deviations from black body could be present. (These however would be much less than those claimed by the American–Japanese team.) Likewise small fluctuations in the intensity or anisotropies should be present. Such perturbations can be estimated depending on which framework or scenario is used to describe the formation of galaxies and structures (see chapter 5).

According to the big bang model, photons, with energies distributed according to the black body spectrum, last interacted with matter at the time of recombination and have remained unaffected since. In fact they could have passed through clumps of cosmic matter and interacted with free electrons present in these clumps. The interaction with these electrons through Compton scattering could have modified the spectrum and broken its purely thermal character.

It is usual practice to introduce the 'Comptonisation parameter' to measure the deviation of the spectrum from black body. To give an example, the (incorrect) result of the American–Japanese team gave a value of $y = 0.03$. The more very hot gas (within which there are many free electrons) in the galaxies and clusters, the more important this process must be. This idea that there might be lots of gas had occasionally been proposed to explain the presence of the cosmic X-ray background. The FIRAS instrument on board the COBE satellite (see below) measured a value of $y < 10^{-3}$. This confirmed that the previous claims of deviations were wrong. It also excluded the presence of large quantities of hot gas in the universe. Thus it is necessary to look elsewhere for an explanation of the X-ray background. Thus it is well established that up to the present precision of measurements, which is excellent thanks to COBE, the CMB obeys the black body radiation law. Thus the thermal character of the radiation and the big bang are spectacularly borne out. Moreover, FIRAS gives the temperature of the CMB very accurately as $T = 2.726\,\text{K}$. However, the actual value is of less interest than the confirmation of the thermal nature of the spectrum.

The COBE and Relikt Satellites

The results that we have just discussed come from the COBE satellite. We shall see a bit later on that it has provided further important results about the CMB. This satellite, constructed at the Goddard Space Center of NASA and launched in 1989, was almost entirely dedicated to observing the CMB. It was equipped with three instruments, DMR, FIRAS and DIRBE.

DMR stands for differential microwave radiometer, and was designed to measure the intensity of the radiation in two different directions in the sky simultaneously and compare them. It could thus make a map of the sky in the microwave band (in the three wavelengths 3.3, 5.7 and 9.6 mm). It was able to detect for the first time, and to a certain extent map out, the anisotropies in the CMB, at least the anisotropies present in regions of the sky separated by more than 7° (see the results below).

FIRAS is a Michelson interferometer, or spectrograph, for measuring very precisely the spectrum of the detected radiation. The spectrum is measured relative to a reference black body on board the satellite. FIRAS gave the temperature of the radiation, and upper limits on the deviations from the black body radiation law. The third instrument, the infrared photometer DIRBE, provided information about the infrared sky, and the contribution made to it from other non-cosmological sources (the contribution from the Milky Way or other galaxies).

The task of the Soviet satellite, Relikt, launched in 1983, was similar to that of the DMR of COBE. The results it obtained are comparable, although less accurate.

Anisotropies in the CMB

One of the most pressing problems in cosmology today is to understand how galaxies and other structures formed, as will be discussed in chapter 5. Today all the possible scenarios predict the existence of small fluctuations in the intensity of the CMB from one region of the sky to another, and these deviations from isotropy can be estimated according to what cosmogenic scenario one takes. Several years ago this question began to puzzle cosmologists. It appeared in fact that the processes invoked to explain the formation of cosmic structures necessarily predicted anisotropies (fluctuations of intensity) at a level that should have been observationally detectable. Although this did not contradict the big bang model, it did militate against the scenarios proposed within the framework of the big bang model for the formation of galaxies. Thus the first scenarios that had been proposed and in vogue for several years were rejected. There were as a result two consequences. Hunting for anisotropies became an international sport: to detect the expected anisotropies or, failing to detect them, to place more stringent limits on these anisotropies. At the same time theoreticians checked their calculations and developed more elaborate scenarios to account for the formation of galaxies.

In fact the situation was not so very dramatic because of the great uncertainties in these scenarios. The processes involved in galaxy formation are so complex it is impossible to calculate with any precision all the expected effects. Nevertheless more and more complicated scenarios were considered, including, for example, the ad hoc introduction of the hypothesis that large quantities of dark matter exists. By giving this matter the right properties it was possible to match observations with theory. However to be obliged to modify a model to make it compatible with observations is never a good sign in science. The observational quest for anisotropies became imperative for the cosmological community, if the disagreement should not become to dramatic. It was in this context that the two space missions, COBE and Relikt, devoted to observing the CMB, were planned. These two missions have now been implemented, and have yielded their results: the first anisotropies have been detected.

The level at which these first anisotropies of the CMB were detected is very weak (less than one part in 10^5). But it is more or less what was predicted by the various scenarios for structure formation currently considered. Having said this, it is still pretty difficult to directly use these results. In fact the anisotropies are differences in radiation intensities between two regions of the sky. The angular separation between these two regions plays an essential part: it can be linked to the spatial dimensions of the phenomena responsible for these anisotropies. For COBE θ is greater than $7°$, which corresponds to spatial scales of several hundred Mpc. (The precise correspondence depends on the values of h and Ω.) This is much greater than the size of a galaxy, or of a cluster, or even a supercluster, and so there can be no direct link between these results and the various scenarios for galaxy formation. It is only by introducing further hypotheses, and fairly strong ones at that (particularly in relation to the scale dependence of the fluctuations), that one can establish such a link. In the last analysis, these results in fact provide little information concerning the formation of galaxies and structure formation.

Their main merit consists in having demonstrated, at a time when people began to give up the hope of finding them, that anisotropies do really exist, that the level of these anisotropies corresponded more or less to what theory had predicted, and that they are detectable. They did not put into question either the big bang, the various cosmogenic scenarios or inflation. (It would be an exaggeration to say that the observations *confirmed* inflation.) These results are still not sufficient to lead to any substantial advance in our

understanding of the cosmos. The search for fluctuations in the CMB is still
not completed.

4.1.8 The dynamical role of radiation

We still have to talk about the dynamical effect of radiation on the universe.
In the same way as matter, radiation has an energy density and therefore
has a gravitational effect, in particular on the expansion law of the universe.
This effect appears in Friedmann's equations. The difference between the
role played by radiation and matter lies in the high pressure of radiation, i.e.
$p = 1/3\rho c^2$ whereas for matter it is negligible. Also the density and pressure
of radiation falls as R^{-4}, faster than that of matter, which falls as R^{-3}.

Today the influence of matter is expressed through the parameter Ω_0.
We do not know its precise value but it is almost certainly between 0.02 and
0.2 (some people believe it might be as high as 1). The effect of radiation
is described by the parameter Φ_0, which is observationally determined from
the energy density of the CMB. The energy density of thermal radiation is
given by the well known formula

$$\rho = a\, T^4,$$

$$a = \frac{8\,\pi^5\,k^4}{15\,h^3\,c^3},$$

where a is the radiation constant and takes the value $7.564 \times 10^{-15}\,\mathrm{erg\,cm^{-3}}$
K^{-4}.

The energy or mass density of the CMB is then $4.4 \times 10^{-34}\,\mathrm{g\,cm^{-3}}$,
which when compared with the critical density, $1.96 \times 10^{-29}h^2\,\mathrm{g\,cm^{-3}}$, gives
$\Phi_0 = 2.25 \times 10^{-5}h^{-2}$.

Ω_0/Φ_0 measures the relative strengths today of the effects due to matter
and radiation. The effect of radiation today is therefore completely negligible. We are in the epoch of matter. On the other hand, the effect of
radiation at any given moment can be written as $\Phi_0(1 + z)^2$ compared with
$\Omega_0(1 + z)$ for matter. Thus they were equal at a time called the matter–
radiation equilibrium*, which is defined by

$$1 + z_{\text{equilibrium}} = \frac{\Omega_0}{\Phi_0},$$

which gives a redshift of around 5000 (see figure 4.1).

For redshifts greater than this, roughly speaking during the primordial
universe, the density of radiation dominated in the evolution of the universe.
We took this into account when we described the evolution of the primordial
universe, where R varies as $t^{\frac{1}{2}}$ (see subsection 3.3.2).

It should be noted, without attaching any particular significance to it, that the periods of matter–radiation equilibrium and decoupling took place at almost the same time. One should also note that at this period the age of the universe is less than ten million years. Therefore, the radiation dominated epoch lasted much less time than the matter dominated epoch. It was this that allowed us to ignore it when we calculated the age of the universe.

In the primordial universe both the matter and radiation density exceeded the value of 2×10^{-21} g cm^{-3}; the temperature (common to both matter and radiation let us remember) was greater than 5000 K. Matter and radiation were distributed very homogeneously. Such are the essential properties of the big bang models confirmed by the remarkable isotropy of the observed CMB.

4.2 Events in the primordial universe

4.2.1 The appearance of structures

All the structures we know on astronomical scales – galaxies, stars, planets, and so on – only came into existence well after recombination. Nevertheless, the history of the primordial universe is marked by remarkable events produced by extreme conditions of density, pressure, temperature and energy. What we are talking about here is essentially the progressive building up of matter as we know it in its microscopic form – production of elementary particles, the synthesis of the lightest atomic nuclei, and above all the establishment of the relative abundances of the different sort of particles, the relative abundances thereafter being conserved and determining the physics of our world. The unfolding of the primordial universe can be reconstituted within the framework of the big bang models.

The universe was then bathed in energetic radiation that was intimately linked to matter and that dominated the dynamics of the universe. No structures existed, whether macroscopic or microscopic. Objects only appeared after recombination.

We have already examined the recombination phase, which marked the end of the primordial universe. Before, there were protons (and helium nuclei) on the one hand and electrons on the other, interacting but separate. Recombination marked the appearance of the elementary structures we call atoms. These could not have existed earlier because the intense radiation at these high temperatures would have destroyed them immediately. Thus it was the cooling of the universe, which itself was linked to the expansion, that allowed the appearance of structure.

In fact, if one goes back sufficiently far into the past of the primordial universe, beyond the period of recombination, atomic nuclei and even the particles from which they are constituted, which can be considered more elementary still, did not exist. They were born in a similar fashion early in cosmic history, by the coming together of still more elementary components.

The production of the lightest atomic nuclei, or what is called primordial nucleosynthesis, is an extremely important stage in the big bang. Before, there were neutrons and protons, which interacted, but which were not bound together. In the process of nucleosynthesis they are brought together to form atomic nuclei such as helium.

The appearance of successive and increasingly elaborate structures are the most characteristic events of primordial cosmic evolution. The latter is marked by various stages related to which type of particles are present. More concretely, if we consider cosmic history after the first second the particles that are known today have already been produced, and the contents of the universe obey physical laws determined by expansion and gravitation on the one hand and by the strong, weak and electromagnetic interactions on the other. These are described by laws that are approximately known, thanks to quantum mechanics and to measurements made using particle accelerators. Matter in the universe behaves like a gas formed of these particles; it is important to consider their abundance and mutual interactions.

4.2.2   Particles in the primordial universe

One of the essential characteristics of the primordial universe, as we have already written, is the more or less perfect thermal equilibrium of its contents. This assumption alone enables us to calculate the properties of the particle populations present; in equilibrium the distribution of each is given by a known formula of quantum statistics. In other words, so far as we have a good understanding of the laws of physics, we know everything about the contents of the universe. The essential question is to know the species of particles present. At very high energies (that is to say right at the beginning of the history of the universe) the theories of particle physics, still poorly understood, should tell us this. When the temperature of the universe decreased, the 'exotic' species of particles disappeared and the population of the universe reduced to the particles that we know today.

These are classified on the one hand as leptons*, and on the other as hadrons. The leptons are grouped into families. We have known three such families for a long time, the electron, muon* and tau*. The number of lepton families, normally denoted by N_ν, which is also the number of neutrino families, was for a long time unknown. (Only the big bang models predicted that it should not be greater than four.) This prediction was only verified in 1989, three months after the new particle accelerator at CERN*, the LEP, went into action. There are three and only three families! In each family there is a lepton – electron, muon or tau – and its antiparticle*, and its associated neutrino (electron, muon and tau neutrinos) and anti-neutrino. Leptons only interact via electromagnetic and weak interactions (as well as gravitational of course).

The hadrons interact through the strong force. Essentially they are baryons (neutrons and protons and their antiparticles) but also pions*. According to the theory of strong interactions (quantum chromodynamics*), the hadrons are made up of what are still more elementary, the quarks*. Thus the really fundamental hadronic particles are in fact the quarks, which interact by exchanging particles called gluons*. Quarks (and antiquarks) are classified into three families corresponding to the three lepton families.

What can one say about the abundance and the distribution of the various species of particles? In the very earliest times in the universe, they were all in mutual equilibrium, which allows us to derive the abundance and distribution of each. (We assume that the chemical potential* of each species is zero.) It is the reactions between all the different species of particle present that make this equilibrium possible. For a given species to be in equilibrium it is necessary that the reactions that are able to couple it to the others proceed sufficiently rapidly. One considers that this equilibrium is maintained so long as the reaction rate at a given time remains greater than the expansion of the universe at that time.

To see whether or not a given species is in equilibrium with the others it is then necessary to study the rate of the reactions that can couple it to the rest of the matter. Very often these rates depend on the density of some other species of particles that participate in the reactions, and that are therefore necessary for this equilibrium. The density of these particles must therefore be sufficient for this coupling.

The case of charged particles is the simplest: at very high temperatures each species is in equilibrium with electromagnetic radiation, even if only by virtue of the process of pair creation and annihilation (a particle can annihilate with its antiparticle to give a photon, as well as the inverse reaction). As far as uncharged particles are concerned (like the neutrinos) it is the weak interactions that assure their equilibrium with the rest of matter, at least if the rate of these reactions is high enough.

In these earliest times, the photons are described, according to the laws of quantum statistics, by Bose–Einstein* statistics, and so long as they are in equilibrium, the other particles (muons, electrons, neutrinos) by the

Fermi–Dirac* distribution. The question is to know up to what moment each species present remains in equilibrium with the rest of the matter.

4.2.3   The neutrinos

The case of the neutrinos is exemplary for several reasons. On the one hand they play a role in the primordial synthesis of helium nuclei: they intervene in the reactions which transform protons into neutrons and vice versa. On the other hand they could be massive and hence play a role in the dynamics of the universe. For this reason much importance has been attached to them recently in cosmology. They represent the archetypal particle that could play a role in the primordial universe and their history can serve as a model which can easily be adapted for any species of particle that could have existed.

Their equilibrium with the other forms of matter can be brought about by the weak interaction. The first question is to ask up to what point in time this equilibrium lasted. The reactions that could couple them to other particles are the following:

$$
\begin{array}{ccc}
e^- + \mu^+ & \longleftrightarrow & \nu_e + \bar{\nu}_\mu \\
\nu_e + \mu^- & \longleftrightarrow & \nu_\mu + e^- \\
\nu_\mu + \mu^+ & \longleftrightarrow & \nu_e + e^+
\end{array}
\tag{4.1}
$$

as well as the reactions obtained by replacing each particle by its antiparticle. The rate of these reactions can be calculated as a function of temperature, density and the abundances of the particle species involved, and will depend on the cross-section for the weak interactions σ_w. To know whether the neutrinos are or are not in equilibrium, it is necessary to work out the rate of these reactions, which of course depends on the abundance of the muons. The problem then comes down to knowing this abundance.

At very high temperatures (around 10^{13} K) the muons are relativistic and in thermal equilibrium with radiation (at the same temperature). Quantum statistics (the Fermi–Dirac distribution applying to muons) tells us that provided they stay relativistic their density, n_{muon}, is of the same order of magnitude as the number density of photons, n_γ. (In fact $n_{\mathrm{muon}} = 3/4 n_\gamma$.) They are relativistic so long as their temperature is greater than $T_{\mathrm{muon}} = 10^{12}$ K, which corresponds to the rest mass energy, mc^2. At T_{muon} the muons become non-relativistic and their abundance falls by the Boltzmann factor* $\exp(-m_{\mathrm{muon}} c^2 / kT)$

Depending on these values, one can show that the reaction rate becomes too low to ensure equilibrium around a temperature of $T = 1.3 \times 10^{11}$ K. Therefore it is at this temperature that the neutrinos decouple.

In fact the electron neutrinos are sensitive to other reactions involving electrons and not muons. Thus

$$e^- + e^+ \longleftrightarrow \nu_e + \bar{\nu}_e$$

$$e^\pm + \nu_e \longleftrightarrow e^\pm + \nu_e$$

(4.2)

In this case it is no longer the abundance of muons that governs the reaction rates but the electron abundance. The decoupling of electron neutrinos therefore takes place later, between T_{muon} and $T_e = 5 \times 10^9$ K.

4.2.4 Other particles

In order to verify whether or not the neutrinos were in equilibrium with other matter, it was necessary to work out the abundance of coupled particles (muons or electrons). To do this we used the fact that these particles themselves were in thermal equilibrium (which, to be fully rigorous, needs to be verified).

The case of the photons is easy to deal with because they have no mass and so remain relativistic. Nor is there any Boltzmann factor to include. Moreover they stay coupled, at least to baryonic matter, throughout the primordial universe. Thus it is useful to use the abundance of photons as a reference.

For any given particle the distribution of momentum, q, is given by the general quantum statistical law

$$n_i(q) \, dq = \frac{4\pi \, h_{\mathrm{Pl}}^{-3} \, g_i \, q^2 \, dq}{e^{E(q)/kT} \pm 1}$$

(4.3)

where h_{Pl} denotes Planck's constant.

The particle is assumed to have mass m (possibly zero, as in the case of the photon) and energy $E(q) = (m^2 c^4 + q^2 c^2)^{\frac{1}{2}}$ and can be a fermion (+ sign in the denominator) or a boson (− sign). Finally the factor g_i corresponds to the number of possible spin states for the particle ($g_i = 2$ for photons, electrons, protons and neutrons). This formula, applied to photons (by putting the mass to zero and the degeneracy factor g_i to 2 and including a negative sign) gives the black body or Planck law.

This formula is very general, but only applicable when the species of particle considered is in thermal equilibrium at the temperature T (and when the chemical potential of the species of particle is zero). It yields, upon integration, the energy density, the particle number density and every other quantity corresponding to each species of particle. For example, the number density of particles, the energy density and the pressure are given respectively by the integrals

$$n_i(T) = \int n_i(q)dq,$$

$$\rho_i(T) = \int E_i(q)n_i(q)dq, \qquad (4.4)$$

$$p_i(T) = \int \frac{q^2}{3E_i(q)} \, n_i(q) \, dq.$$

These equations can be applied to each species of particle and their antiparticle. It follows (in so far as the chemical potentials are zero as we have supposed) that the abundances of a given particle and its antiparticle are the same.

So long as the temperature is considerably greater than the mass of the particle in question, the formulae for the different particles does not vary much from particle species to particle species. In fact, at high temperatures, the mass is negligibly small in the energy term, and so the latter is identical for all particle species. Only the factor g_i is important, and the bosonic or fermionic nature of the particle. According to whether one is interested in bosons or fermions, the number density of particles is given by

$$n_B = \frac{g_B}{2} n_\gamma,$$

$$n_F = \frac{3g_F}{8} n_\gamma$$

and their energy density by

$$\rho_B = \frac{g_B}{2} \rho_\gamma,$$

$$\rho_F = \frac{7g_F}{16} \rho_\gamma,$$

(where the subscript γ signifies photons, F fermions and B baryons). Of course this only holds so long as kT is greater than the rest mass of the particle considered, or in other words the particles are relativistic.

When this is no longer the case, one can use the approximation that abundance is given by the relativistic value above multiplied by the Boltzmann factor $\exp(-mc^2/kT)$.

It is expedient to take the abundance of photons as a reference, and to define the abundance of protons relative to photons, or whichever species of particle relative to the photon abundance. We shall see later the advantage of doing this. We have already studied this phenomenon in the case of muons, whose abundance fell when the temperature dropped below $T_{muon} = 10^{12}$ K, which brought about the decoupling of the muon neutrinos. We shall illustrate it below in subsection 4.2.6 with the example of the electrons.

4.2.5 From quarks to nucleons

If we are to describe the history of the primordial universe with a degree of confidence, we cannot go too far back in time and high in temperatures since the nature of the particles present and the laws of physics become too uncertain.

According to the current theory of the strong interaction (quantum chromodynamics), the nucleons are themselves made up of still more elementary particles, the quarks. In the 'very primordial' universe, the nucleons were not yet formed, and the universe must have been filled with quarks. It is only around a temperature of T_{QH} (about 200 MeV) that the quarks combined with one another to form pions and nucleons. It is convenient to call this the quark–hadron transition*. This transition marks the beginning of the 'hadronic epoch'. We are relatively unable to describe the physics that held sway before this transition. It is only later that the universe can be described in terms of known physical laws.

At this epoch the universe contained, in addition to the light particles, a large number of relativistic pions, in fact almost as many as the number of photons. The nucleons (protons and neutrons), which were in equilibrium with the other matter, had ceased to be relativistic. They and their antiparticles were much less abundant than the photons.

The end of the hadronic epoch is defined by the disappearance of the pions. This occurs when they become non-relativistic at a temperature of around $kT_{pion} = 130$ MeV $= 10^{12}$ K, that is at 10^{-5} s after the big bang. Their abundance falls with the Boltzmann factor $\exp(-m_{pion}c^2/kT)$ and becomes very small.

After the disappearance of the pions, the abundance of all the hadrons – pions and nucleons – is much lower than the abundance of photons. This is the end of the hadronic epoch, and the beginning of the leptonic epoch.

4.2.6 Story of the electrons

The mass of the electron is 0.511 MeV, and corresponds to a temperature of $T_e = 5 \times 10^9$ K. As long as the temperature of the universe remains above T_e the abundance of electrons (and positrons) is almost the same as that of photons:

$$n(e^-) \approx n(e^+) \approx n_\gamma.$$

The above follows, as we have already noted, from the fact that one can neglect the mass of the electrons and positrons in the number density that one obtains from the Fermi–Dirac law. The latter becomes almost indistinguishable from the Bose–Einstein law expressing the abundance of photons. (We note in passing that n_γ decreases as $R^{-3}(t)$.)

However, as T approaches T_e, the mass term becomes important in the Fermi–Dirac formula, and the electron abundance decreases with the Boltzmann factor $\exp(-m_e\,c^2/kT)$. The physical process responsible for this disappearance is electron–positron annihilation producing photon pairs. So long as the temperature is greater than T_e, these annihilations are compensated for by the inverse reaction, electron–positron pair creation. Of course this inverse reaction can only take place when the pairs of photons have energy greater than the mass of the particles produced in the reaction, say 0.5 MeV. Thus once the temperature of the radiation lies below T_e the number of sufficiently energetic photons is relatively so low that the rate of this reaction becomes negligible. Electron–positron annihilation is thus no longer compensated for by the inverse reaction of pair creation, and the abundance of electrons and positrons falls irreversibly.

These annihilations themselves later cease when the electrons and positrons become too rarified to undergo collisions owing to the dilution caused by the expansion of the universe. Their relative abundances (relative to the photons) thus freezes out at these current values.

4.3   Primordial nucleosynthesis

During the first three minutes of its existence, the universe only contained individual particles, and no structures such as atomic nuclei. Essentially in the form of matter there were neutrons, protons, and electrons, and in the form of radiation photons and neutrinos (assuming the latter to have zero mass). There were no bound states of several nucleons.

However, in these favourable conditions that existed during the big bang, neutrons and protons fused to form light atomic nuclei. ('Light nuclei' contain only a few nucleons, i.e. protons and neutrons, as opposed to 'heavy nuclei such as carbon, nitrogen, and oxygen etc., which contain many.) Conditions favourable to such nucleosynthesis existed several minutes after the Planck era at a temperature of around 10^9 K. Isolated nucleons were brought together to form the nuclei of deuterium*, helium, and lithium*.

Astronomers observe these elements in the universe with the abundances predicted by these models of primordial nucleosynthesis. Furthermore, there is no other explanation of these abundances since these elements cannot have been formed in the interiors of stars in the same way as the heavier elements such as carbon, oxygen or iron. The big bang models are in this instance once again vindicated.

4.3.1 Nucleosynthesis – the old story

In the beginning the big bang models were developed in the hope of explaining the abundance of all the chemical elements in the universe, from the lightest (hydrogen and helium) up to the heavier elements (such as carbon, nitrogen, oxygen, the metals etc.). Astronomers had not yet understood that stars were in fact effective nuclear factories for the production of the chemical elements. It was suggested that early in the history of the universe conditions existed under which these elements could have been formed. In other words matter had been hot and dense.

In fact, if the density and the temperature had been sufficiently high for the production of heavy elements corresponding to their present day abundances, then far too much of the light elements would have been produced. The process could not have worked. Moreover, the universe of today is no longer dense and hot. It must therefore have cooled during its expansion. This expansion must be taken into account in the calculations. It became clear that expansion played an important role, and for the whole theory to work it was absolutely necessary. Eventually the model became more specific: heavy elements were not synthesised during the primordial universe, but much later and in an entirely different way within the stars. Only the light elements were formed in the primordial phase. The big bang models came into being.

Only deuterium D, helium ^{3}He and ^{4}He, and small amounts of berylium ^{7}Be and lithium ^{7}Li were synthesised in the primordial universe, at a temperature of around one thousand million kelvin and by successive capture of neutrons from primordial matter made up of protons, neutrons and electrons. The essential reason why the reactions could not continue up to the synthesis of heavier nuclei is the absence of any stable nuclei with $A = 5$ and $A = 8$ (recall that A is the atomic mass and is defined as the total number of nucleons, i.e. protons and neutrons, in the nucleus).

Very quickly adjustments made to the model gave a prediction that fossil radiation should exist at several kelvin. Scientists started to search for this radiation in 1964, and it was found the following year by radio astronomers.

Fine tuning

In fact the processes are very ordered, and it is almost miraculous that the parameters are so precisely adjusted. Given this fine tuning the model lays down quite severe constraints.

Everything depends on the fusion reaction between protons and neutrons to form deuterium:

$$n + p \longrightarrow D + \gamma$$

The binding energy of the deuterium is 2.23 MeV, corresponding to a temperature, T_D, of 2.58×10^{10} K. It is this reaction that is responsible for the luminosity of the majority of stars. So long as the temperature of the universe is greater than this temperature deuterium nuclei cannot survive. It is the initial production of these nuclei that begins the build up of heavier elements according to the reactions

$$D + p \longrightarrow {}^3He + \gamma$$
$$^3He + {}^3He \longrightarrow {}^4He + 2p. \tag{4.5}$$

4.3.2   The formation of deuterium

Since the above element conditions the whole chain of production, it is necessary to know how much is formed. For temperatures above 10^9 K, the species present, protons, neutrons, electrons and positrons and neutrinos, are linked together by a number of reactions governed by the weak interaction. These reactions are more or less balanced.

$$p + \bar{\nu}_e \longleftrightarrow n + e^+,$$
$$n + \nu_e \longleftrightarrow p + e^-, \tag{4.6}$$
$$n \longleftrightarrow p + e^- + \bar{\nu}_e$$

During these phases, neutrinos and electrons along with their antiparticles are in thermal equilibrium, and their abundances which are governed by the Fermi–Dirac distribution depend on the temperature. From knowledge of these it is possible to calculate the rate of the various reactions, and ultimately to calculate the parallel evolution of neutrons and protons as a function of cosmic time.

First of all it is necessary to evaluate the production rate of D, the first step in the process. If it is sufficient (greater than the rate of expansion of the universe) the equilibrium of the reaction is guaranteed. The proportion of deuterium produced as a function of time depends on the abundances X_p and X_n of protons and neutrons, as well as on the temperature and pressure. The binding energy of the deuterium nucleus plays of course a crucial role in the calculation. The amount of deuterium produced remains very small so long as the temperature does not drop below the critical temperature T_D. Above this temperature the balanced reactions favour neutrons and protons to the detriment of the helium nuclei. Furthermore, the energetic thermal photons break up the few deuterium nuclei already formed by photodisintegration. Thus the fabrication of the subsequent elements is blocked. The production of deuterium (which itself depends on the condition $T < T_D$) thus constitutes a bottle neck for primordial nucleosynthesis.

On the other hand, once the temperature drops to T_D, all the available neutrons are rapidly incorporated into deuterium nuclei. Of these nuclei, only a small fraction avoid transformation into helium during the following step. Everything in fact is dependent on the number of neutrons present.

4.3.3 The abundance of neutrons and protons

In the primordial universe, neutrons and protons are transformed into one another via the reactions written above. The relative abundance of the two species depends on the rate at which the corresponding reactions take place. The conditions are different according to whether or not the temperature of the universe is greater than $Q = (m_n - m_p)c^2 = 1.293$ MeV or in other terms 1.5×10^{10} K, the mass difference between the two types of particles. So long as $T > Q$, one can calculate the rates of the reactions (in both directions p→n and n→p) from the law governing the weak interactions* depending on the relative abundances of particles involved. They are equal and significantly less than the rate of expansion of the universe. This assures equilibrium between the protons and the neutrons, so that the ratio of their abundances is given by the Boltzmann law

$$\frac{N_n}{N_p} = e^{-Q/kT}.$$

Nevertheless, the reactions between the two species slow down because of the expansion and the consequent dilution. At a temperature of $T_{n/p}$ their rates fall below the rate of expansion: the ratio N_n/N_p moves away from equilibrium behaviour and freezes at a value of around $\exp(-Q/kT_{n/p})$. For example, if $T_{n/p} = 10^9$ K, the ratio is fixed at around 0.15 to 0.2. Below $T_{n/p}$ (at a time $t_{n/p}$ of about 20 seconds) the reactions are too scarce to maintain equilibrium.

Following this the decay of free neutrons (the lifetime of a neutron is around 900 seconds) further reduces the abundance of neutrons exponentially in time. Once the formation of deuterium has been made possible (that is after t_D when the temperature has fallen below T_D), this happens very quickly and more or less completely. All the neutrons present now exist in the form of deuterium.

4.3.4 Helium production

One can roughly estimated the amount of ^{4}He produced by assuming that all the neutrons available at the beginning, and which have first gone to form

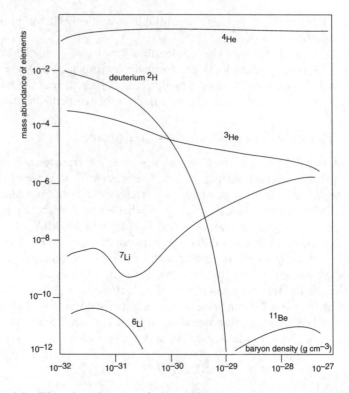

Figure 4.4: The abundances of the various chemical elements predicted by models of nucleosynthesis as a function of the present mean baryonic density of the universe. The shaded rectangles show the observed values. (After Zel'dovich and Novikov, *The Structure and Evolution of the Universe*, University of Chicago Press, 1983.)

deuterium, end up in the form of ^{4}He nuclei. This of course depends on one neglecting the occasional deuterium nuclei that escape helium formation, but in this approximation it is fully justified. We shall accept that all the neutrons present at time t_D have been incorporated into helium nuclei. With the knowledge that a helium nucleus contains two neutrons, one can deduce the mass abundance of helium, which turns out to be about 25%.

More detailed calculations that take into account the whole ensemble of reactions and the various chemical species that are involved allow one to precisely calculate the helium abundance and furthermore the abundance of deuterium that has escaped fusion and remains still in its free form. All these reactions are fairly delicately tuned, and the resulting abundances are sensitive to a number of parameters (see figure 4.4).

Usually one chooses the first of these parameters to be the ratio, η, of the number of baryons to the number of photons:

$$\eta = 2.83 \times \Omega_{\text{baryons}} h^2 10^{-8}.$$

This quantity, which has not changed during the cosmic history following primordial nucleosynthesis, expresses the velocity of expansion of the universe at the time of nucleosynthesis. It constrains the abundance of helium produced. In fact, once all the neutrons have been incorporated into deuterium, the crucial question is to know how much of the latter escaped being transformed into helium. The greater the velocity of expansion (the smaller η), the greater the number that escape and the higher the present deuterium abundance must be. Observations to date imply $\eta < 10^{-9}$. Simultaneous compatibility with the abundances of all the elements imposes the constraint

$$3 \times 10^{-10} < \eta < 5 \times 10^{-10}.$$

This translates into a constraint on the baryon density parameter of the universe:

$$0.01 < \Omega_{\text{baryons}} h^2 < 0.018.$$

This constraint is extremely important, since it gives an upper limit to the baryon density (about one fifth of the critical value), which rules out a universe that is both flat ($\Omega = 1$) and totally made up of baryons (at least if one excludes more complex scenarios for nucleosynthesis).

There is another parameter that plays an important part. We have already seen that the neutron to proton ratio freezes out at a temperature of $T_{\text{n/p}}$. Its value at the start of nucleosynthesis depended on this temperature, as does the total amount of helium produced. It is through this temperature that the amount of helium produced depends on N_v, the number of neutrino species that exist. (Roughly speaking the amount of helium produced increases by one percent for every extra family of neutrino.) The observed abundances impose a limit of $N_v < 5$, which was for a long time the best known limit on N_v. Today experiments carried out at LEP in Geneva have shown that $N_v = 3$, which one can take as further confirmation of the predictions made by the big bang models.

4.4 The very early universe

4.4.1 The particle universe

If there were no atoms before recombination, it was because the intense energy that existed at that time (in the form of radiation for example) would have immediately broken up any atom present. By the same token, before nucleosynthesis radiation would have destroyed any deuterium nucleus. As one goes back in time the energy continues to rise, to such an extent that even elementary particles themselves cannot exist. During the first microsecond after the big bang the universe was filled by a 'soup of quarks' (the quarks are as far as we know the 'ultimate' constituents of matter). In the course of the quark–hadron transition, these quarks formed into neutrons and protons. But the physics before this transition is still uncertain. Today's theories are incapable of describing the first microsecond of the universe as the energy and density are so high. It is in the realm of speculation.

The very early universe

The first microsecond in the history of the universe is undoubtedly rich in events, but everything depends on the theory adopted to describe the interactions between particles at very high energies and densities. There is no properly developed theory, only provisional ones. Physicists are particularly interested in grand unified theories* or GUTs, which are a special case of what are called gauge theories. It is remarkable that GUTs, even without being completely specified, do make a number of cosmological predictions. There was an explosion in new cosmological ideas related to GUTs in the period from 1970 to 1980 – inflation, the possible existence of magnetic monopoles, and cosmic strings. These notions and their consequences have been abundantly discussed. No doubt this interest was partly a question of fashion, but there is still a lot of interest in them if only because they have shown that the big bang models can give rise to new and interesting possibilities.

The essential aspects of these predictions derive from the fact that the GUTs imply the existence of a 'phase transition' in the first few fractions of the first second of the universe. More specifically, this occurs at the moment when the GUTs are no longer applicable, and need to be replaced by theories more in keeping with the physics holding today. In order to understand this transition, called symmetry breaking*, it is necessary to study a few properties of the GUTs.

4.4.2   The unification of interactions

Most physical processes can be described in terms of four types of interaction (electromagnetic, gravitational, the strong, responsible for the binding of the nucleus, and the weak which is involved in beta decay). We have known since the last century that electricity and magnetism, apparently different, are two aspects of the same phenomenon, electromagnetism. Is it possible that other interactions, apparently different, might be unified in the same way?

An affirmative answer appeared thanks to the concept of a gauge theory*: each interaction can be considered as a consequence of a symmetry, called a gauge symmetry, that operates in an internal space (not the usual geometric space) associated with the properties of the particles. These abstract notions can be used to describe each of the different interactions, barring gravitation.

They do however go further. The physicists Glashow, Salam and Weinberg constructed a theory, called the electroweak* theory, describing simultaneously and in a unified manner electromagnetism and the weak interaction.* From this arose the idea that perhaps from a more complete symmetry it would be possible to construct a grand unified theory or GUT* in which the strong interaction also was incorporated into a common scheme.

Physicists have not yet been successful in constructing such a theory, but most believe that it will happen soon, and feel that they are able even now to make some statements about its properties. These are sufficient to make certain cosmological predictions.

In this way, one fundamental property of the GUTs is that they predict two distinct regimes in physics. At a high temperature of say 10^{28} K and corresponding to mean particle energies of 10^{15} GeV, the physics of 'grand unification' holds sway, or more properly, the interactions lose their distinctions and a great number of particles coexist. At lower temperatures, in the other regime, the symmetry of the theory is broken, and the various interactions become distinct and physics reduces to the physics that we know today.

4.4.3 Phase transitions

In the big bang model the temperature of the universe does exceed, during the first 10^{-35} s, a value corresponding to this limit of 10^{15} GeV. Grand unification must therefore have held. In cooling down the universe ends up in a phase described by ordinary physics. In a way a bit like the sudden transformation of water into ice, there would have been a phase transition corresponding to this symmetry breaking.

This episode in cosmic history has some remarkable consequences. In fact this transition from grand unification to ordinary physics did not take place identically everywhere in the universe. Although it happened perfectly coherently in any given region, in different regions it would have taken place differently. It is somewhat analogous to the surface of a lake that begins to freeze at different points forming large homogeneous zones of ice that do not fit together perfectly. This comes about because the physical state

transition zone

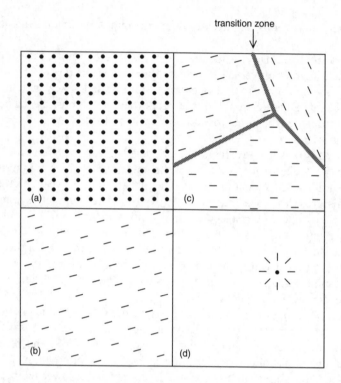

Figure 4.5: The different configurations of the quantum field vacuum can be schematically represented by the orientation of a vector. This can be zero everywhere (a), have a constant orientation (b), or be constant in certain regions (c). The last case can give rise to transition zones or topological defects, such as magnetic monopoles.

after the transition, although it behaves in accord with the ordinary laws of physics, is not strictly determined by theory and can slightly differ from one point to another.

The structure of grand unified theories is fairly complex. However, one can visualise, by analogy, certain aspects of the phase transition by following the length and direction of a characteristic vector (see figure 4.5). The domain of grand unification corresponds to the simplest situation: the vector is everywhere zero, as represented by the points in (a) in figure (4.5). This illustrates the symmetry of the theory, since there is no preferred direction. At low energy this symmetry is broken, and the vector is oriented differently at every point. The case where the orientation is the same at every point corresponds to ordinary physics, although the size and orientation of the vector has no influence. In this very simple configuration (b) of figure (4.5) the symmetry is broken since a privileged direction has emerged. However, there is no particular physical manifestation since the vector does not vary from one point to the next. This case corresponds to the vacuum state of ordinary physics. By contrast, when the orientation from one point to the next does vary, there will be perceptible effects. Moreover, cosmological models predict that the vector can only be constant in spatially limited regions, that is within domains of coherence. Within each of these domains, the vector stays the same, but it varies rapidly at the interfaces of these zones (c).

Thus after the transition the universe will be partitioned into different zones. In each ordinary physics holds. But they do not mesh together perfectly and leave defects at the junctures – borders where physics has not been able to become ordinary physics and where vast quantities of energy are trapped. According to variants of the grand unified theories, these defects should occupy surfaces (domain walls), lines (cosmic strings), or point defects (magnetic monopoles). These regions that have escaped symmetry breaking are called generically 'topological defects'. Within them reposes energy (and therefore mass) in a very concentrated form. Depending on the case, the cosmological implications will be different.

4.4.4   Magnetic monopoles and cosmic strings

The case of point defects is remarkable because they can be identified with a type of object whose existence was foreseen within a very different context. This is the magnetic monopole* (see figure 4.5). It would appear from the evidence to date that it is impossible to separate the north pole of a magnet from the south pole: apparently an isolated magnetic charge or 'monopole' does not exist. However, the point defects that result from the phase transition in GUTs can be identified with such an object. These would be microscopically small and almost point like regions, within which one can consider the physics of grand unification to be trapped.

Such configurations store up lots of energy, so that they appear as super massive particles (almost a microgram). In fact these are really regions where space itself is different rather than elementary particles (in the sense of an electron for example). According to theory the interior of such a region would be structured like an onion, with physics further and further removed from ordinary physics and closer to grand unification applying to successive layers as one penetrates deeper (with a characteristic length scale of 10^{-28} cm). Nevertheless, the details of this internal structure are masked from us since probing the interior of the magnetic monopole would require enormous energies. Viewed from outside, that is on a scale of 10^{-13} cm, it would seem like a particle.

It is possible to roughly and somewhat naively predict how many magnetic monopoles would have been created in the phase transition period. The universe was only 10^{-35} s old. There would have been no physical process that would have been able to act beyond the dimension of the horizon at that epoch, which would have been around 10^{-35} light-seconds (or about 10^{-25} cm). Hence the domain of coherence (in which the phase transition could have taken place uniformly) could not have been greater than this size. One would thus expect about one magnetic monopole for each domain of coherence, which would amount to one in each volume of 10^{-75} cm^3. But these objects are so massive that the mass density calculated from the numbers above would be completely incompatible with cosmological observations. The universe would contain so much mass, in the form of magnetic monopoles, that it would have collapsed a long time ago under its own weight!

Either the grand unified theories are not valid, or these magnetic monopoles have suffered a drastic dilution. Without going into the details, we should note that the 'inflationary' (see later) cosmological models do propose just such a dilution, and therefore make the existence of primordial magnetic monopoles compatible with cosmology. This process would have been so effective that only a few magnetic monopoles would have been left in the universe. Therefore the GUTs predict either too many monopoles, or hardly any. Unfortunately experiments and observations will probably not give us any answer to this question for a long time.

Their detection is not however impossible, but will require new experimental techniques. Since they are very heavy, they will experience very little loss through ionisation and will hardly ever be stopped by matter. Furthermore the expected flux is very weak, and so very high performance detectors are required. Despite the various attempts, no confirmed detection has taken place. Were such a detection to occur it would force a modification of electromagnetism, and would (ultimately) bear out the grand unified theories. Undoubtedly it would be a major event in physics of this century.

Cosmic strings

According to other versions of the theory, symmetry breaking in grand unification could engender string like entities with enormous energy densities. These are called cosmic strings and manifest themselves essentially because of their mass, which is distributed along immense filaments produced at the time of symmetry breaking. They could be the explanation of the formation, several thousand million years later, of the first galaxies. Even if these objects could in principle produce observational (gravitational) effects, their existence must still be considered a highly speculative, albeit fruitful, hypothesis.

4.4.5 Vacuum energy and inflation

GUTs have introduced another important contribution, the concept of cosmic inflation. In fact the latter is not specifically linked to the GUTs (indeed it now seems almost impossible to construct an inflationary model within the framework of the GUTs). Quantum field theory* (which provides the general framework for the discussion of GUTs and other theories) predicts in effect, particularly in the conditions of the primordial universe, the possible presence of a form of energy, the vacuum energy, associated neither with radiation nor particles. It is possible to foresee some of the properties this form of energy would necessarily have, which could be described, somewhat formally, by an equation of state admitting negative pressures, $p = -\rho c^2$ (which would be used in Friedmann's equations to give a contribution analogous to that of the cosmological constant).

If this form of energy dominated the universe during some part of its history, then it would have had a very special influence on its dynamics. In place of the relatively moderate expansion (in which the scale factor varies as a power law of time), during such an inflationary period the expansion is very rapid and exponential. During such a small time period as 10^{-35} s all cosmic dimensions would be amplified by a factor of say 10^{50}, which is comparable to the effects of expansion accumulated over the remaining period of thousands of millions of years!

Cosmologists have incorporated this process into their models of the universe, and have realised that on the whole the consequent effects help with the predictions of these models – on the one hand through the extreme dilution that takes place, and on the other from the modifications that it brings about in the causal structure of spacetime. Apart from certain points in relation to dark matter and galaxy formation, it appears to help answer a number of questions in cosmology. A thorough examination of the compatibility of the existence of such an inflationary period with the big bang models and particle physics constitutes today a very active and lively field of research. Even so it seems difficult to establish a coherent inflationary scenario and it is possible that this idea will be abandoned as quickly as it was adopted.

4.4.6   The big bang revisited

It is still to early to know whether such innovations will be recognised and accepted by the majority of astrophysicists. No doubt the theories of particle physics developed in the coming years will bring about new ideas, and just as original. In any case, particle physics has played a fruitful role during the last few years. One instance of this is the suggestion of the possible presence of a massive particle which, even though practically undetectable, would have important dynamical effects and which could provide a solution to the problem of dark matter and galaxy formation. On the other hand particle physicists have come to see the early universe as an extremely favourable field for their theories, and one from which they can obtain constraints on their models. In any case, the concepts introduced in this instance can be seen as fertile and original. Particle physics has caused our notions in relation to the evolution of the cosmos to evolve. To know whether or not these models will be able to withstand their own implications is still not possible.

Physical theories have been developed to describe physics at higher and higher energies and temperatures, and are now capable of addressing the questions posed at earlier and earlier times in the evolution of the cosmos. Nevertheless there does remain one fundamental hurdle to overcome, and this is the crucial step. The way to reconstruct the very first instants remains blocked by the lack of any genuine synthesis between the theories of gravitation (general relativity) and quantum mechanics.

5

Galaxy formation

5.1 Gravitation and fluctuations

5.1.1 The fundamental question

The problem of galaxy formation, or more generally the formation of large scale cosmic structures, remains one of the most delicate questions of cosmology, and appears to resist all analysis. There is no model today that is able to satisfactorily answer all aspects of the problem.

Does this mean that there are still aspects of the big bang models that need to be revised? Up to now it seems as though we have been able to answer each question taken separately. It is only when we try to put these answers together and place them in the unified framework of the big bang that things go wrong. No scenario is today capable of explaining all the various stages of galaxy formation.

The essential question is how to understand the appearance of inhomogeneities as important as the stars, galaxies and clusters of galaxies whilst the universe was in the past so very homogeneous, as the big bang models suppose. How can we reconcile the initial homogeneity as assumed by these models (and apparently confirmed by the measurements of the CMB), with the appearance of the universe today? If this continues to be impossible, we shall have to find another model to describe the universe. It will be up to this model (no doubt less homogeneous to begin with) to explain the great isotropy of the CMB.

Even if there is no generally accepted scenario, most accept that the fundamental mechanism for galaxy formation is 'gravitational instability'. Density fluctuations would have been created early in cosmic history, and then amplified and concentrated under self-gravity up to the point where the structures we observe today were formed. In all probability this process was also aided by physical processes other than gravitational ones. Thus it is necessary to understand where these fluctuations came from, and how they were able to grow and condense.

Using dynamical and physical laws (see subsection 5.2.2) we can make an elementary calculation of the evolution of these fluctuations and estimate how fast they grow. Here we come to the first fundamental problem: if we know the present level of the fluctuations (deduced from observations of galaxies, clusters etc.) we can retrospectively estimate the level necessary in the history of the universe in order to obtain the level of fluctuations today. In particular we can estimate the level of the fluctuations at the time of recombination.

Furthermore, we have a means of measuring this level. Since the background radiation (CMB) observed today has propagated without alteration since that epoch, it must carry with it the imprint of all fluctuations present at that time. However, the analysis of these observations shows these traces to be extremely weak. The same must therefore be true of the density fluctuations. This is the major contradiction, and the one that the majority of models come up against.

5.1.2   Gravitational instability

The basic principle is that the universe was very homogeneous on all scales. To question this 'cosmological principle'* underpinning the big bang models would be to put into question the models themselves and their essential foundations. This would go against the ground rules that astrophysicists, cosmologists and scientists in general accept. Before abandoning a class of models that have in other ways been so useful, one should do one's utmost to reconcile the model with reality. (Besides, what can one replace it with?)

The hypothesis of homogeneity allows one to describe the global evolution of the universe in terms of its average density and pressure, ρ_0 and p_0, as we showed in an earlier section. The history of the formation of galaxies and other structures in the universe looks at the deviations from this homogeneity. These deviations can be expressed at each point in the form

$$\rho(\mathbf{x}) = \rho_0 + \delta\rho(\mathbf{x})$$

This defines the density fluctuation, $\delta\rho(\mathbf{x})$, and the relative density fluctuation or density contrast

$$\delta(\mathbf{x}) = \frac{\delta\rho(\mathbf{x})}{\rho_0}$$

Most calculations concerned with galaxy formation use $\delta(\mathbf{x})$, the density contrast field. One practical difficulty comes from the fact that the fluctuations condense simultaneously on all scales – stars form, galaxies condense. The galaxies group together in clusters and superclusters. These phenomena, which are more or less simultaneous, act on one another. Even without talking of intermediate scales, the situation is very complex.

In order to make some progress, astrophysicists make a fundamental approximation: that fluctuations on different scales can be studied independently, at least during a part of cosmic evolution. This approximation is certainly justified so long as the relative fluctuations (measured in terms of δ) have not become too great. There is no consensus concerning the limiting value of δ below which this approximation is still valid. This linear approximation requires δ to be less than 1, but this limit is certainly not restrictive enough. What limiting value of δ should be adopted and even whether this approximation is valid in the context of the standard models is still a subject of debate.

In any case, it is almost certain that the linear approximation can be applied on all scales up to a certain epoch (which itself will depend on the various scenarios). Certainly this will be the case for redshifts greater than 100, and probably for redshifts greater than 10 and may be for still lower redshifts. We shall call this the linear epoch, in contrast to the non-linear epoch that followed up to the present day.

The linear epoch must be divided into stages, for reasons that will be clear later. The crucial period is recombination and we shall see that the nature of the fluctuations completely changes at this moment.

We shall also see that whether the evolution is linear or non-linear depends in fact on the scale considered. According to most models, the development of the fluctuations obeys the linear approximation for certain scales (the largest), but not for others.

Linear growth

The linear phase stretches from the beginning of cosmic history (in fact from the first appearance of 'primordial' fluctuations) up to at least $z = 100$. The question is to know the growth in the level of these fluctuations. This problem can be tackled in the same way as most problems in physics for which the initial conditions are given (the appearance of the primordial fluctuations). These are characterised by a mean fluctuation level, which is necessarily very weak (the mean value of δ is less than 10^{-4}). These fluctuations then evolve in accord with physical laws up to some final state. The problem is to know this final state.

There is a necessary condition which all the models must fulfil, and which results in an enormous constraint. This fundamental constraint is the formation of galaxies. Since today the density fluctuations exceed the linear domain (on the scale of galaxies and galaxy clusters δ is considerably greater than 1), the linear phase must have grown until it was no longer valid. In other words during the linear phase the density contrast must have grown from its initial value of less than 10^{-4} up to the critical value of around 1. It is remarkable that this constraint does not depend on the following evolution, during the non-linear phase, but only on the development of the linear phase, which is much simpler to follow. One supposes in the first analysis that the non-linear phase, which is much more difficult to study, is sufficiently effective that once started it results quickly in the formation of galaxies. Almost all the models that have been proposed more or less accept this. Without this assumption, the constraint would be all the more severe.

The first question to know then is whether during the linear phase the level of fluctuations could have achieved a value close to 1. Gravitation is the driving force behind the amplification of these condensations, and the growth in the factor δ. It has to offset the expansion, which has the opposite effect of diluting matter more and more, and to overcome pressure which also conteracts any contraction. The history of the formation of galaxies is to a large degree the history of this competition.

5.1.3 Fluctuation scales

In fact it would be too simplistic to study the evolution of only the mean value of δ. It would also be superfluous (and impossible) to study the evolution of $\delta(\mathbf{x})$ at all points of the universe. But at the very least it is necessary to separate the various scales. This separation is made fairly easily during the linear phase by decomposing the fluctuation field into its contributions to different scales, much as one separates light into its spectral components.

The idea is that each scale that is present will result at some given time in the formation of a corresponding structure. Each scale can be characterised by and associated with a typical mass (in units of the solar mass): 10^6 for star clusters, 10^{11} for galaxies, 10^{12} for clusters of galaxies, and 10^{14} for superclusters etc. It is also useful to identify each fluctuation scale with the mass that it contains.

What is the spatial scale or size associated with a given fluctuation? If there has been no condensation, the matter contained in the fluctuation will remain associated with a constant comoving volume, or in other words, with a constant comoving length, and thus with an actual length that varies with $R(t)$. The presence of a fluctuation makes the local expansion slightly slower than the expansion for the universe as a whole. Nevertheless, during the linear phase the difference in the rates of expansion is very small. In the first approximation one can consider the characteristic size of the fluctuation to follow the cosmic expansion and hence simply vary as $R(t)$, the scale factor. In other words, one can identify a fluctuation with a constant comoving length, $l_{comoving}$, which is related to the actual length by the relation

$$\frac{l_{comoving}}{l} = 1 + z.$$

Once the linear phase has come to an end, the matter content of a fluctuation deviates more and more from the expansion law, and the containing volume grows less and less quickly in comparison with the cosmic expansion. It attains a maximum radius (this moment is termed the turn-around) and finally starts to shrink, heralding actual condensation. As a result, the size of a condensation of a galaxy, say, is much smaller than the associated fluctuation. The latter is, by definition, represented by the radius of a sphere that would have contained the mass of a galaxy had there been no condensation. It is roughly equal to the mean separation distance between galaxies today, of around 1 Mpc, which is at least 10 times the diameter of a typical galaxy.

5.1.4 The statistics of the fluctuations

Since it is impossible to describe all the characteristics of the density field, one resorts to a statistical description in terms of probabilities and mean values. One can introduce, for example, the probability $P[\rho(\mathbf{x}_1),\ \rho(\mathbf{x}_2),\ldots\rho(\mathbf{x}_i)]$ that the density takes values $\rho(\mathbf{x}_1)$, $\rho(\mathbf{x}_2),\ldots\rho(\mathbf{x}_i)$ at the points $\mathbf{x}_1$, $\mathbf{x}_2\ldots\mathbf{x}_i$. However, this description is too detailed.

The fluctuation spectrum

The density field $\rho(\mathbf{x})$, or rather the density contrast, $\delta(\mathbf{x})$, can be expanded into Fourier components, $\delta(\mathbf{k})$. $\mathbf{k}$ is a wave vector whose magnitude, k, identifies a scale length $l = 1/k$. The Fourier amplitudes, $\delta(\mathbf{k})$, are complex quantities. The strength (magnitude) of the $\mathbf{k}$ mode characterises the level of the fluctuations corresponding to $l = 1/k$, and one defines the power spectrum of the fluctuations as the quantity $P(\mathbf{k}) = \langle|\delta(\mathbf{k})|^2\rangle$, which in words is the mean square fluctuation as a function of the reciprocal scale length.

This power spectrum is a fundamental quantity. In effect, if the statistics are not too complicated, it provides almost all the properties of the density fluctuation field. For example, the mean value of the fluctuations in mass of a volume of fixed size and form is given by

$$\int_0^\infty P(\mathbf{k}) \, W(\mathbf{k}) \, d^3\mathbf{k},$$ (5.1)

where $W(\mathbf{k})$ is a window function chosen to single out the properties of the volume being studied.

Often one is interested in the velocity field corresponding to these fluctuations. This can also be developed into Fourier modes. The velocity is assumed to have zero vorticity, so that the Fourier modes only have wave vectors parallel to the velocity, viz.

$$\mathbf{v}(\mathbf{k}) = \frac{\mathbf{k}}{k} v(\mathbf{k})$$

In the same way one can define the power spectrum of the velocity field $P_v(\mathbf{k}) = \langle v^2(\mathbf{k}) \rangle$.

The power law spectrum

In galaxy formation one of the aims in carrying out the calculations is to find the fluctuations as a function of time. We shall see that during the linear phase these changes can be calculated.

One of the first questions to address is the behaviour of the initial spectrum, which results from the processes connected with the primordial fluctuations. Unfortunately the latter are unknown. In the absence of any consensus over this point it would seem reasonable to leave its form free and parametrise it, for example, by a simple power law. It is customary to write the spectrum in the form

$$P(\mathbf{k}) = A k^n$$

(A different spectrum could be decomposed into pieces having this form with different values of n.) To guarantee convergence, the values of n cannot be chosen too large or too small. The value $n = 1$ represents a special case where the spectrum is 'scale invariant', and is also called the Harrison–Zel'dovich spectrum. Apart from its simplicity, it corresponds to the particularly interesting case where the potential (gravitational) energy is equally distributed at all scales.

It should be noted that the constant A is never specified *ab initio*, but is always adjusted by normalising a posteriori from observations.

Knowledge of the spectrum allows one to calculate the mean value of the density fluctuations or the mass within a sphere of comoving radius, L

say, containing a mean mass of M. Thus

$$\frac{\delta M}{M} \propto \frac{\delta \rho}{\rho} \propto L^{-(3+n)/2} \propto M^{-(3+n)/6}.$$

Often one defines the index $\alpha = (3 + n)/6$, so that the above equation becomes

$$\frac{\delta M}{M} \propto \frac{\delta \rho}{\rho} \propto L^{-3\alpha} \propto M^{-\alpha}.$$

The fruitfulness of this approach derives from the fact that fluctuations on different scales (different modes) evolve independently of each other. This is true for all fluctuations up to the non-linear regime* (around $z = 100$). This could still be true today for larger scales (above about 8 Mpc) where fluctuations are still in their linear phase. These we shall call the linear scales.

5.1.5 The connection between statistics and objects

To understand the formation of galaxies, we have to know how to go from the density fluctuations to the formation of objects. Even if one is not able to follow the development of the physical processes, an operational model must be able to say how one should calculate the distribution and the properties of any given astronomical object (galaxies for example) from the density spectrum.

The most commonly used methods involve several different steps. First it is necessary to choose the scale, either of a galaxy or a cluster, depending on the type of object being studied. One then smooths the field, in order to remove the fluctuations on small scales that are deemed to have little influence on the formation of the objects under study. This smoothing is carried out using a convolution with a window function as mentioned above.

The resulting field only contains fluctuations at (and above) the scale chosen. It is then necessary to apply some more precise prescription: for example that all the regions of the universe where the relative density is greater than some level correspond to the presence of galaxies (or whatever objects are of interest). This somewhat arbitrary procedure (although it could be refined and modified) rests on a very strong and little justified hypothesis. Nevertheless it seems to give acceptable results and, in the (hopefully provisional) absence of any better understanding of the non-linear condensation, it provides a reasonable first approach. However, it would of course be pointless to hope to definitely confirm or invalidate a model on the basis of such procedures, although many astrophysicists appear to forget this. Despite its shaky foundations, this procedure (called the 'Press and Schechter' method) is used a lot. The distribution of galaxies, for example, according to this method, comes about from the distribution of the 'maxima' of the density field of the universe.

Nevertheless none of these models succeeds in reproducing the observations. A supplementary condition (albeit reasonable) has been introduced: the distribution of the density field (and of the peaks) does not directly reproduce the distribution of the matter that one can observe. On the one hand, the 'visible' objects, such as galaxies, clusters etc., only represent a part of the contents of the universe (this is the idea behind dark matter cf. subsection 3.2.5). But above all the distribution of visible matter does not follow the distribution of this dark matter. Nevertheless the models, in order to maintain a degree of simplicity, and so that they do not become unwieldy, assume a not too complex relation between the value of the density field at a point and the probability that a luminous object (galaxy) appears there. This is an extreme simplification of a poorly understood situation. The corresponding models go under the name 'biased structure formation'. The simplest version supposes that objects (such as galaxies) form when the value of the field of density fluctuations is greater than or equal to b times the mean fluctuation (b is called the bias parameter). More complicated versions of the theory suppose that b can vary with scale. Others appeal to different laws. So long as one forgets the physics, everything is a priori permissible, and one simply introduces further parameters to 'save appearances'. Of course all this only goes to show that our present models of galaxy formation are far from satisfactory.

5.2   The growth of fluctuations

5.2.1   Stages in the growth of fluctuations

The various models of galaxy formation can all be schematised into the following stages:

creation of the initial fluctuations,

relativistic growth (before recombination),

freeze out or damping,

recombination,

post-recombination linear growth,

non-linear growth,

collapse of objects.

The essential differences in the various scenarios proposed for galaxy formation depend on the behaviour during the first stages, which have repercussions for the subsequent evolution.

The period of recombination is crucial and it is convenient to begin our discussion there. Given the level of the fluctuations then, the consequent evolution starts off by being linear. Let us first study this period of evolution. Later we shall return to the earlier period, and ask ourselves how the initial fluctuations were created and how they evolved up to recombination.

Linear condensation is a competition between three effects: gravitation, which brings about the condensation of fluctuations under self-gravity; expansion, which in contrast tends to dilute these fluctuations; and gas pressure, which opposes contraction. Remember that during the period following recombination radiation is decoupled dynamically from the gas, so that the dynamics of the latter can be studied independently.

The basic assumption in the linear calculation, and what makes it possible to solve the equations, is that δ is small and less than 1. In this approximation the evolution at different scales is independent. (Undoubtedly this is the most questionable aspect of the calculation.) The smallness of the level of fluctuations simplifies the calculation and makes it fairly straightforward. We shall adopt the simple and currently adopted view that this approximation is applicable. The operative principle in the various scenarios for galaxy formation is that the fluctuations are created with small values of δ, which subsequently grow linearly.

5.2.2 Linear growth and the Jeans mass

It might seem strange to begin the problem of the growth of the fluctuations with the intermediate stage following recombination. However, this stage constrains all the models. Its study requires a hydrodynamic calculation involving gravitation (of the gas condensations acting on themselves) and pressure, all within the context of the general expansion. These calculations show that gravitation and condensation can only win out if certain conditions are fulfilled. These conditions can be expressed by what is called the Jeans criterion.

The greater the mass contained in a fluctuation, the greater the importance of gravitation compared with pressure. For small fluctuations pressure wins, and the fluctuations cannot condense. For larger scales, on the other hand, gravitation dominates and the fluctuations condense. The critical value that separates the two regimes is called the Jeans length (or the Jeans mass if one prefers to describe the fluctuations by their mass). If the size of the fluctuation is exactly equal to the Jeans length, then the effect of pressure exactly equals that of gravitation.

At recombination, given the temperature of the gas (about 4000 K) we can obtain its pressure, and find that the Jeans mass has a value of around 10^6 solar masses. This value has a special importance. Recombination marks the start of the real growth of baryonic fluctuations. All fluctuations on a scale greater than this begin to condense, whilst the others remain constant. Hence one can see the Jeans scale as a lower limit to the scale of structures which will form by this mechanism. All we now need to do is to calculate how fast the condensation takes place. Calculation shows that the rate of growth is more or less the same on all scales greater than the Jeans mass.

The condensation of fluctuations

Were it not for expansion, this growth would be very quick (exponential) and the galaxies (and other structures) would form very rapidly. However, expansion has the effect of strongly reducing the effectiveness of gravitational instability. In fact the level of fluctuations increases (at most) in proportion to the scale factor, $R(t) = R_0/(1 + z)$, in other words roughly as time t to the power $\frac{2}{3}$. Since recombination takes place around $z = 1500$, this means that a given fluctuation level at time t is given by

$$\delta = \frac{\delta_{rec}(1 + z_{rec})}{1 + z} \tag{5.2}$$

where δ_{rec} denotes the level of the fluctuation at recombination. This calculation only remains valid so long as the linear approximation is valid, that is up to the instant z_{nl}, which depends on the scale being considered, where the perturbation becomes non-linear. (In the simplest version this is when δ gets close to 1.)

Applying this formula to the period stretching from recombination to the onset of non-linearity, z_{nl}, we have

$$\delta_{nl} = \frac{\delta_{rec}(1 + z_{rec})}{1 + z_{nl}}$$

If we replace δ_{nl} by the value 1 in this formula, and using the fact that $z_{nl} > 0$, we find immediately that

$$\delta_{rec} > (1 + z_{rec})^{-1} > 7 \times 10^{-4}$$

Observations of the CMB indicate in the first analysis that the mean level of fluctuations at this epoch was smaller than this value. In order to resolve this apparent contradiction, it is necessary to carry out more detailed calculations and introduce further assumptions. One must also study more closely the fluctuations before recombination and their effect (at recombination) on the CMB.

Moreover, the fluctuations condense proportionally to $R(t)$ only at the beginning of the linear phase, depending on the value of Ω. In fact it remains valid up to about $z = \frac{2}{5}\Omega$. For $z < \frac{2}{5}\Omega$, the (linear) growth of the fluctuations practically ceases. This can only further constrain the level of fluctuations at recombination.

5.2.3   Fluctuations before recombination

As far as we know there are three possibilities for the generation of the first fluctuations. Two of them are connected with a phase transition in the primordial universe (the inflationary model or cosmic strings at an epoch of around 10^{-35} s after the big bang). The other assumes that the fluctuations date from the Planck time (10^{-43} s). This is tantamount to saying that the fluctuations come about because of the 'initial conditions' of the big bang models. Since physics is still unable to say anything whatsoever about these initial conditions, it would be presumptuous to insist that the initial conditions responsible for the fluctuations have a character more or less 'natural' or 'probable' than any other. Whatever the process, we can suppose that the fluctuations were created at an instant very close to $t = 0$.

Remember that in the primordial universe, the density of radiation by and large dominates matter by a factor greater than $z/10\,000$. It is necessary to know to what extent radiation also undergoes fluctuations. Two types of fluctuations, which do not have the same consequences, should be distinguished.

For modes of the first type, the fluctuations in radiation are such that the total density is unperturbed. One can therefore write

$$\delta\rho = \delta(\rho_{\text{matter}} + \rho_{\text{radiation}}) = 0.$$

However since

$$\rho_{\text{radiation}} \gg \rho_{\text{matter}},$$

it is clear that $\delta\rho/\rho$ is much smaller for radiation than for matter. The relative perturbation of the radiation density is almost zero, whence the term (strictly speaking inexact) isothermal is given to these fluctuations. However, if the total density is unperturbed, then these fluctuations bring about no change in the gravitational potential, nor, for that matter, of the (local) curvature of the universe. They are therefore called isocurvature fluctuations.

They are also sometimes called entropy fluctuations. Fluctuations due to the presence of cosmic strings would be of this nature. Concentrations of matter produced by the presence of cosmic strings would be off-set by a relative absence of radiation at the same place.

By contrast, adiabatic or *isentropic* fluctuations do not change the ratio of matter density to radiation density, so that

$$\frac{\delta\rho_{matter}}{\rho_{matter}} = \frac{3}{4} \times \frac{\delta\rho_{radiation}}{\rho_{radiation}}.$$

The generation of adiabatic fluctuations are predicted by the inflationary models.

The initial spectrum

It is also necessary to have a knowledge of the statistics of these fluctuations. In the absence of any defined process for their generation, arguments of simplicity or generality suggest that they are Gaussian. Inflationary models certainly favour this. Models based on cosmic strings, on the other hand, predict a different statistical behaviour based on objects (cosmic strings) that have already condensed and yield fluctuation statistics that are not Gaussian. We shall not however enter into the details here.

In general the spectrum is assumed to follow a power law as defined in subsection 5.1.4. Simplicity and generality suggest a scale invariant power law which corresponds to $n = 1$ (see subsection 5.1.4), and this is mostly adopted in the models. Primordial fluctuations generated by a process linked to a phase transition appear to give in a fairly natural way this scale invariant spectrum.

Finally, in order to know the fate of these fluctuations, we have to specify which matter components they involve. If the universe was made up uniquely of massive baryons, there would be no problem. This was the situation in the standard big bang models and the first scenarios proposed for galaxy formation where matter was supposed to be purely baryonic. More recently difficulties with purely baryonic models have led one to consider the possible existence of a large quantity of matter in the form of non-baryonic particles which do not interact (except gravitationally) either with other matter and radiation, or mutually. This has the effect of modifying the scenarios for galaxy formation. As we shall see, we have to distinguish between cold, warm, and hot particles.

5.2.4 Relativistic growth

Once the fluctuations have been created we can calculate their evolution, which will depend on the properties of these fluctuations. One cannot use the Newtonian approximation for gravitation since the density is so high that even for relatively small fluctuations the fluctuations in density are sufficiently high to produce a considerable local curvature. Thus a relativistic treatment is essential.

If we carry out the relativistic calculations it is natural to introduce a Jeans length and mass which depends on the radiation pressure, since matter and radiation are strongly coupled before recombination. Before recombination, and so long as the universe was dominated by radiation (before the moment of matter–radiation balance), the Jeans length can be calculated in terms of the speed of sound for radiation, which is equal to $c/\sqrt{3}$. The Jeans length is very large, and almost equal to ct, that is virtually the same size as the horizon of the universe at the same instant. It grows in proportion to time, that is as $(1 + z)^{-2}$. (It does not vary in proportion to the scale factor.) It follows easily that during the period before recombination and matter–radiation balance, the Jeans mass (in the same way as the baryonic mass contained within the horizon) varies as $(1 + z)^{-3}$.

Fluctuations can condense so long as they are on a scale greater than the Jeans length (mass). For example, for a fluctuation on the scale of a galaxy (about 10^{11} solar masses), the mass stays above the Jeans mass for about a year after the formation of the fluctuation. (In fact one year after the big bang the mass contained within the horizon is equal to 10^{11} solar masses.) It thus condenses during the first year of cosmic history, just at the time when it 'enters the horizon'. One sees immediately that the larger its size the longer the time a fluctuation will take to condense.

A relativistic calculation yields the speed of condensation. For an adiabatic fluctuation, the density contrast δ grows more or less linearly in time t. The total growth of an adiabatic fluctuation for a given mass before recombination is therefore proportional to the time during which it is greater than the Jeans length. The greater the mass of the fluctuation, the longer the time interval.

In contrast for an isocurvature fluctuation, the level remains effectively constant from its origin to the time of recombination.

Primordial growth of the spectrum

Evidently this process modifies the behaviour of the spectrum, since amplification differs according to the scale considered. Let us suppose that the initial spectrum obeys a power law of the form $P \propto k^n$, which translates into a mass fluctuation law of the form

$$\delta M / M_{\text{initial}} \propto \delta \rho / \rho_{\text{initial}} \propto L^{-(3+n)/2} \propto M^{-(3+n)/6}.$$

We define a transfer function $T(L)$ as the level of amplification experienced by the scale L. For an adiabatic fluctuation this level of amplification

is proportional to the time the fluctuation remains outside the horizon: a
calculation shows that a fluctuation of comoving length L is amplified by
a factor proportional to L^2. In other words, $T(L)$ is proportional to L^2 (or
$M^{\frac{3}{2}}$). However, for scales greater than about 10^{17} solar masses (defined as
the Jeans mass at the event – recombination or matter–radiation balance –
occurring first). The fluctuations are always greater than the horizon and
therefore greater than the Jeans mass. They never enter the horizon, and
therefore never stop growing. The transfer function does not change for all
scales greater than this value.

One should therefore expect a change in slope of the spectrum at this
scale length. For isocurvature fluctuations, which do not experience any
amplification, the transfer function stays equal to 1 regardless of the scale.

In order to obtain the form of the spectrum at recombination, we
simply multiply the initial spectrum by the transfer function. For adiabatic
fluctuations, large size fluctuations are very effectively amplified and the
spectrum is augmented for these large scales. We should remember that
we have never talked about the absolute level of amplification but only the
relative level for different scales.

5.2.5   The recombination spectrum

The preceding calculation concerned the primordial phase of amplification
of the fluctuations between the time of formation until the moment when
they enter the horizon. However, once the fluctuation has entered the
horizon matters are not finished (right up to recombination). If some (for
example isothermal baryonic fluctuations) stay as they are with no changes,
frozen until recombination, others experience damping, which has the effect
of annulling the previous growth.

Isothermal baryonic fluctuations are not damped. Imprinted on, or
'trapped against', a uniform background of radiation, they alone remain
unchanged up to the point of recombination. Radiation, for its part,
maintains its homogeneity up to recombination.

But other types of fluctuations are damped. This damping is important
as it is responsible for the differences in the scenarios for galaxy formation.

The damping of adiabatic fluctuations

Once the adiabatic baryonic fluctuations (which we should remember involve
both baryons and radiation more are less equally) enter the horizon, they
stop condensing. Instead they start to oscillate like acoustic waves which
effect both baryons and photons. However, electron scattering of photons
damps these oscillations of the fluid, which consists of gas and radiation,

particularly for the smallest scales. An analysis of this damping shows that fluctuations on a scale less than some limit, called the 'Silk length (mass)', completely disappear. This limit, depending on the cosmological parameters, is about 10^{14} solar masses.

The evolution of the spectrum

Once we know the initial conditions and phases in the history of the fluctuations before recombination, we can calculate the characteristics, namely the spectrum, at that time. (Later we shall examine the fluctuations involving particles other than baryons and photons.) There are at least two reasons why these characteristics are particularly important. The first is that radiation at the present epoch bears their imprint. Some of the properties of this radiation are preserved unchanged from this period until today, and so should still be recognisable from observations of the CMB, notably measurements of the angular variations. In principle this should allow one to test the different scenarios. Secondly, the instant of recombination plays the role of initial conditions for the second matter dominated phase of cosmic history in which fluctuations begin to grow linearly and end up in the condensation of galaxies and the formation of other cosmic structures.

The spectrum at recombination is equal to the initial spectrum multiplied by the transfer function $T(L)$. After this, however, the adiabatic baryonic fluctuations undergo damping, as described above. The spectrum is thus equal to the initial spectrum multiplied by $T(L)$ but with a cut-off (reduction to zero) at scales smaller than the damping mass.

5.2.6 Connections with the cosmic background radiation

These calculations make it possible to predict the spectrum at recombination given the initial spectrum and initial conditions. Nevertheless, since there is no prescription for determining the initial conditions it would be an illusion to hope to obtain constraints on the conditions at recombination. In fact the inverse process should be invoked: given the constraints that observations of the CMB impose on the fluctuations at recombination, we should be able to deduce constraints on the initial conditions, and hence on the processes giving rise to the primordial fluctuations. It turns out that these constraints are not strict enough to favour any particular process for producing the initial fluctuations.

As far as the models are concerned, observations of the CMB provide us with a means of reasoning retrospectively as explained above: given the constraint that objects have already condensed, and that fluctuations have therefore gone non-linear, and given the maximum rate of linear amplification from calculations, we can predict the minimum level necessary at recombination. This minimum level must imply a minimum level in the fluctuations of the CMB, which could be compared with observations. These calculations are fairly complex and involve somewhat detailed interaction processes between matter (baryons) and radiation, as well as the effect of fluctuations in the gravitational potential on the latter. In principle this can tell us the expected statistics of the angular fluctuations of the CMB depending on the framework of the different models of galaxy formation. This strong constraint virtually eliminates adiabatic baryonic scenarios.

The role of non-baryonic matter

If the universe is dominated by something other than ordinary matter things are different. From the point of view of the mass, it is in fact possible for the main components of the universe not to be baryons but some other non-baryonic particles that do not interact, or at least only very weakly, either with themselves or with radiation. This hypothesis of non-baryonic dark matter seems fairly gratuitous, but there is at present no observation that contradicts it.

From the perspective of galaxy formation, it has a double role. On the one hand it allows a dense universe, possibly with the critical density, without violating any of the constraints imposed from primordial nucleosynthesis (see section 4.3) which only apply to baryonic matter. On the other hand, the non-baryonic matter envisaged does not couple to electromagnetic radiation (this was the reason for inventing it). Thus at the moment of recombination the fluctuations of this non-baryonic matter might have been present without producing any of the effects of baryonic matter, and hence without violating constraints from the isotropy of the CMB.

As a consequence, the fluctuations in the gravitational potential resulting from these non-baryonic fluctuations would have helped the baryonic fluctuations to condense. The latter would then have been able to grow up to the point where they formed condensed objects. Globally condensation of baryonic matter would take place more easily without any contradiction with the observations of the CMB. The general idea here is welcome; the problem is to find a scenario that would coherently describe the successive stages of the process. There is presently no such scenario that is completely satisfactory. Even if the idea seems seductive it hardly works any better, when the numbers are put in, than the more simple and more economic (in the sense that there are fewer ad hoc assumptions) scenarios.

One of the difficulties with the non-baryonic matter scenarios is that here too the fluctuations are damped, although for different reasons from the baryonic case. As with baryonic matter, we define a mass cut-off scale in the spectrum. In fact one classes the various types of particle likely to be important into three categories according to the value of this cut-off scale. For hot particles, the damping scale is of the same order of magnitude as the Silk mass (corresponding to the mass of a cluster or supercluster of galaxies). At least qualitatively the corresponding models are similar to those based on adiabatic baryonic fluctuations.

On the other hand, if the cut-off mass is sufficiently small that one can neglect it, as in the case of 'cold' particles, damping is irrelevant to those scales that interest us (above 10^6 solar masses). These scenarios are similar to those with isothermal baryonic fluctuations.

Finally, one can define an intermediate case in which the particles are very naturally called 'warm'. Here the damping mass is of the same order of magnitude as the galaxies, so that the spectrum is cut off precisely at these scales.

5.2.7 After recombination

After the electrons recombine with the ions, baryonic matter becomes neutral, and no longer interacts with radiation. Once free from the effects of radiation, matter is able to contract under its own weight in a purely baryonic universe, or failing this, aided by the presence of fluctuations of non-baryonic matter. So begins the linear phase of growth studied in subsection (5.2.2): all fluctuations grow until their scale is greater than the Jeans mass (say, at the time of recombination, 10^6 solar masses). They do this at roughly the same speed, with δ increasing as the scale factor $R(t)$. (This holds when the fluctuations are purely baryonic; if non-baryonic fluctuations are present the process is more effective.)

Thus the spectrum grows and conserves the form it had before recombination. The first question concerns the moment when any given scale of fluctuation goes non-linear, $\delta = 1$. This calculation is simple. Since the amplification varies as

$$R \propto \frac{1}{1+z}$$

the growth in the spectrum follows the law

$$\delta = \delta_{\text{rec}} \left(\frac{R}{R_{\text{rec}}} \right) = \delta_{\text{rec}} \left(\frac{1 + z_{\text{rec}}}{1 + z} \right),$$

which is valid so long as growth is linear, that is to say, so long as the density contrast is less than 1. It is generally assumed that if some scales (the smallest in most scenarios) have reached the stage of non-linear contraction, the larger ones continue to evolve according to this linear law.

The important question to know is whether the fluctuations on the scale of galaxies have had time, since recombination, to reach the non-linear level, and if so at what moment. If the model is to be valid, this must not happen too late (and so for values of z that are not too small) for galaxy and galaxy cluster scales, since we do observe these objects at redshifts greater than 1.

Non-linearity and dissipation

We have now to consider what happens after the fluctuation becomes non-linear. Unfortunately, all calculations of the non-linear growth are impossible except in a few ideal cases, such as spherically symmetric condensation of a fluctuation. Numerical simulations of gravitational condensation can serve as a guide, but it is still impossible to establish firm results.

The situation is complicated by the fact that other process than gravitation intervene, at least on the galaxy scale. In condensing, matter heats up and radiates energy. This loss of energy is compensated for by a loss in potential energy, which accelerates condensation. The solution of this problem is therefore not something which will be achieved today, and to test the different models of galaxy formation is still difficult.

5.2.8   Doubts and uncertainties

It is necessary to draw as many conclusions as possible from the linear calculations. It is usual, for example, to calculate the 'mass function' (which gives the fraction of objects that have condensed as a function of their mass) using the linear model. The predictions from this calculation agree well with observations. Should we see this as a simple coincidence? Or should we really believe that the complex and violent phenomenon of galaxy formation is partially determined by initial conditions.

As we have already said, the firmest constraint is that fluctuations corresponding to condensed objects should have already reached or surpassed the non-linear scale. But this poses the question at what scale non-linear phenomena start to be important. If this limit is usually taken to be given by $\delta = 1$ it should be borne in mind that the behaviour of gravitational plasmas is complex and it has not been shown that non-linear phenomena, and the interaction between different scales, do not come into play at smaller values of δ. Moreover, all the arguments are based on the mean values of δ, whilst in reality the distribution of its values is poorly understood, and could have a more complex form. Thus a certain scepticism is called for.

As for as the constraints arising from the observations of the CMB, we need to know whether it is the trace of fluctuations at recombination (which is not necessarily the case if there was a late recombination). Finally the observations on large scales, where it would seem reasonable to assume fluctuations are linear, appear to pose a problem. On the one hand there is considerable observational uncertainty. But worse, the results we have seem to be in disagreement with all models, and even perhaps with their most basic assumptions (the role of gravitation, the statistics of the fluctuations and perhaps with the cosmological models themselves).

The distribution of galaxies on non-linear scales (less than 8 Mpc) and their individual properties are beginning to be fairly well known. Unfortunately, since we are still unable to study the non-linear, and hence dissipative, development of the fluctuations, these data are of marginal use.

Other scenarios?

It is difficult to imagine other scenarios. Gravitation seems to be preponderant everywhere we observe in the universe. Besides, what other process could act on such scales?

However, some models invoke another 'engine' for structure formation: the effect of pressure forces (of gas or matter) or primordial turbulence.

An 'explosive' scenario for example assumes that after recombination cosmic explosions might have been able to sweep matter up, causing it to condense and form structures. But one still needs to understand the nature of these explosions (giant stars, quasars?). And of course these models must also satisfy all the constraints cited so far. So far there seems to be no solution that resolves more questions than it poses.

5.2.9 The appearance of galaxies

When did the first galaxies appear? And what did they look like? We still do not know the answer to these questions. A galaxy comes into existence when the first stars light up. Before this happens, what does a mass condensation on the point of collapse, i.e. a protogalaxy, look like? Such an object cannot contain stars and must therefore be difficult to observe. Even so there are some clouds that have been observed already that might be protogalaxies.

Nor do we know what a galaxy that has just formed, or primordial galaxy, would look like. It must be luminous, but we have no idea of its brightness, its colour, or the spectral range in which it emits preferentially. These properties are very probably quite different from those of any galaxy observed today. It would depend on the mass and type of stars formed at this stage, on the amount of dust present, and on the possible excitation and heating processes for the gas etc. No doubt we would not be able to definitively identify such an object even if we did find it in the field of view of a telescope. Some recent observations have uncovered gas condensations and stars that could be galaxies in the process of forming.

We have no knowledge either of the role played by collisions between galaxies and protogalaxies in the formation process. Conventional scenarios only involve fluctuations that are supposed to have condensed more or less individually, but astronomers know very well that clouds interact and fuse, or fragment. This takes place as by virtue of non-linear gravitational interactions and dissipation.

We are really unable to follow these physical processes, and for this reason unable to predict the formation of galaxies. Some galaxies observed in infrared (in particular by IRAS) reveal significant star formation. No doubt this is set off by tidal effects due to the proximity of a neighbouring galaxy. Galaxy encounters may bring about more dramatic events such as the fusion of two galaxies and transformation of spiral galaxies into lenticulars or ellipticals. This sort of interaction is evidently expected to happen mainly in regions where galaxies are abundant such as in the centres of rich galaxy clusters. Giant elliptical galaxies, called cD galaxies, have often swallowed several galaxies.

The distribution of galaxies

The processes forming galaxies must also have been responsible for their distribution in space. The observation of the latter must provide a test of the models. If the general ideas we have presented are right, the linear domain (scales above about 8 Mpc) gives us information about the initial conditions of the process. Effectively on these scales only linear effects have been important, and they are assumed to have to a large extent preserved the 'state of things'.

On the other hand the non-linear distribution has experienced an important and complex evolution which could not have (unless proved otherwise) conserved the initial properties. It is thus pointless to hope to directly test the initial properties or the predictions of the linear theory from a study of non-linear scales. On the other hand, these scales should tell us about the dynamics of these non-linear processes.

The most favourable situation is undoubtedly between the boundaries of the two domains, that is at about 8 Mpc. This is the smallest scale on which we can hope that the linear predictions might be valid. It is also the largest scale on which the statistics are well enough established by observation. It seems as though the density contrast has a value precisely in the region of 1 (or at least for the visible matter component) used in general to normalise the calculations. In fact so far we have only spoken about spectra, that is about the relative fluctuations, but never about the absolute level, which it still seems out of the question to predict. Since one has to have an absolute value for comparison with observations, it is enough to fix the spectrum at some scale (of normalisation) and it is this scale of 8 Mpc that is generally chosen.

Non-linear results

In the non-linear regime, the measured correlation functions (defined in subsection 1.2.4) appear to be in rough agreement with a power law

$$\xi(r) = \left(\frac{r}{r_0}\right)^{-\gamma}$$

where $\gamma = 1.7$ and the correlation length, r_0, is between 5 and 7 h^{-1} Mpc (r is the separation between galaxies). There are two interesting things about this formula. First of all the measured level provides a normalisation to the spectrum, as indicated above. Further the power law form of this function can be interpreted as a scaling law.

Higher order correlation functions (only the three and four point correlation functions are measurable) appear to be expressible in a simple way as functions of the two point correlation function, and are consistent with a similar scaling law. This, together with other results from the statistical indicators, goes to suggest that the distribution obeys certain scaling relations. This new fact has without doubt not yet received sufficient attention. We do not understand how it can be explained by dynamics, yet it is also probably the most important result relating to the non-linear domain. It is likely that the usual formalism is poorly adapted to dealing with this property.

Properties at large scales should be easier to interpret, but are still poorly known. The distribution of clusters of galaxies and superclusters seem to obey similar scaling laws to those of the galaxies. Quasars, on the other hand, seem to be distributed more or less at random. However, the incompleteness of the observations does not allow one to draw any definitive conclusions. Nor do we really understand the nature of these objects. At the present time it is difficult to place them within the framework of galaxy formation models.

In the same way the presence of huge empty regions or voids in the cosmos has important consequences; the corresponding fluctuation level almost attains the value 1 at these large and reputedly linear scales. Can this distinction between linear and non-linear still be maintained? Further more, the level of anisotropies this must produce in the CMB is somewhat difficult to reconcile with the observations at angular scales of 10 minutes of arc to 1 degree!

6

Conclusions

If the models that we have presented provide a coherent vision of the universe and one that is in relative agreement with observations and physical principles, there still remain several problems of a diverse nature.

In the first place, astronomical observations still have not reached the level where we can map out the universe on scales where it truly appears homogeneous. Galaxies are grouped together in clusters, and these in superclusters. Furthermore these objects, with elongated or flattened forms and separated by large empty zones, seem to form larger structures. It is only at larger scales still that homogeneity might exist. Astrophysicists do not doubt that the conclusions drawn from the study of homogeneous models can be applied to the universe, which is filled with these structures.

However, we still do not understand how these structures were formed in the universe, which was assumed to be so homogeneous in the beginning. How could these inhomogeneities have arisen and then grown to make up structures as dense and diverse as the galaxies and clusters, stars, planets, filaments and walls etc. At the present time the weak point of our cosmology is the formation of galaxies.

Equally there are a number of questions of principle that arise concerning the density of the universe, the cosmological constant, the dominance of matter over antimatter, and many others. These are not really questions that the models should resolve, but rather epistemological questions concerning these models themselves. We cannot demand that the big bang models find solutions that are outside their field of application! The recent literature seems to show that there are some people who have not understood this.

In any event, it is remarkable that the big bang models explain the observations so well and are so coherent. Moreover, it would be extremely difficult to find at the present time other models that would be able to give

127

a description of the universe in agreement with astronomical observations. They succeed infinitely better than any other model. This situation might be transitory. Will new observations confirm these models, or will they definitively show them to be wrong?

Glossary

Note: starred words here and in the text refer to words which are explained in the glossary.

Andromeda: name of a constellation. It is also the name of the closest spiral galaxy to our own.

antiparticle: with each type of particle there is an associated antiparticle of the same mass, but with the opposite electric charge (or in fact any other sort of charge). The behaviour of an antiparticle is exactly analogous to that of the particle, and this translates into an important physical symmetry. When a particle collides with its corresponding antiparticle, the two annihilate to give photons. (The inverse reaction can also take place.) However, the universe appears to be filled with particles, and almost free of antiparticles. The reason for this is poorly understood, and this is referred to as the antimatter problem.

baryon, baryonic: baryonic matter is made up of baryons, that is to say, of protons and neutrons, or of nuclei of atoms which are made up of protons and neutrons. All visible matter in the universe is baryonic. It is not excluded that non-baryonic matter might exist in an unknown and invisible form.

black body: this is an idealised object in physics assumed to be in perfect equilibrium with electromagnetic equilibrium. The properties of the latter, and in particular its spectrum, referred to as thermal, are given by the laws of quantum statistics* (the Bose–Einstein* formula) and only depend on the temperature.

Boltzmann factor: see quantum statistics.

Bose–Einstein: see quantum statistics.

boson: see quantum statistics.

CERN: Centre Européen de Recherches Nucléaires.

chemical potential: this is a thermodynamic variable of the same type as

129

temperature. When there is a mixture of chemical species (or particles) in equilibrium, in order to work out the state of the system one needs to know the chemical potential of each species. In cosmology one assumes, in the first instance, that the chemical potential of all species are zero, although interesting consequences follow from different hypotheses.

correlation function (CF): This is a function that provides a means of fixing certain statistical properties of a field or a distribution of objects. It takes on higher values the more the distribution departs from a purely random distribution, and the more the objects are clustered together.

cosmic strings (see also symmetry breaking): these are extremely dense and elongated physical objects which could have been created very early in the history of the universe. They would have in this case probably been at the origin of the first fluctuations which much later initiated galaxy formation.

cosmological principle: this states that the universe must appear the same whatever the point from which it is observed. It implies that the universe is homogeneous and isotropic and forms the foundations of all of present day cosmology.

curvature: see manifold.

dark matter: apart from the massive objects we can observe and identify, there are several indications that some matter could also exist in an invisible form. This is called dark matter. We do not know its composition (possibly non-baryonic*) or how it is distributed.

declination: angular coordinate used to fix the position of the stars in the sky perpendicularly to the equator.

decoupling: this is the instant in the history of the universe where matter (baryons and electrons) and electromagnetic radiation have ceased to be coupled. After this point they evolve independently. According to standard chronology this instant took place one million years after the big bang, when the temperature was about 4000 K.

deuterium: chemical isotope of hydrogen. Its nucleus consists of a neutron and proton. Cosmic deuterium was formed during primordial nucleosynthesis*.

distance indicator: one can never know directly the distance of a galaxy or of a distant astronomical object. To estimate it one measures the value of a physical quantity that one has already shown to be correlated with distance. For more and more distant objects, one uses different distance indicators.

Their succession defines what is called the distance ladder.

distance ladder: see distance indicator.

electroweak: see weak interactions.

equation of state: equation linking the pressure of a fluid to its density.

Euclidean: 'ordinary' space has zero curvature; its properties are described by what is called Euclidean geometry. The space is called Euclidean.

Fermi–Dirac: see quantum statistics.

fermion: see quantum statistics.

gauge theory: a special category of quantum field theory* based on symmetry principles. Quantum chromodynamics*, electroweak*, and grand unified theories* are examples.

general relativity (GR): the theory of gravitation announced by Einstein in 1916. It superseded Newton's theory of gravitation. It provides the foundations for modern cosmology.

geodesics: the geodesics are special curves defined on the manifold*. Trajectories of bodies in free fall (e.g. those of galaxies in the universe) and those of light rays, are geodesics in spacetime.

gluons: see quantum chromodynamics.

grand unification, grand unified theories (GUTs): GUTs are gauge* theories which attempt to simultaneously describe electromagnetic, and weak and strong nuclear interactions. Although the idea is very attractive, to date there exists no satisfactory version of this theory.

hadrons: these are particles that experience strong interactions (see quantum chromodynamics).

helium: the lightest element after hydrogen. Its nucleus is made up of two protons and two neutrons for helium 4 (^{4}He), the most common isotope. The nucleus of helium 3 (^{3}He) only has one neutron. Most helium present in the universe was synthesised during primordial nucleosynthesis.

Hubble parameter: the Hubble parameter or Hubble constant, H_0, defines the rate of cosmic expansion. The 'recession speed', V, of an object situated at a distance D is given by $V = H_0 D$ (Hubble's law). It is the logarithmic derivative of the scale factor*.

Hubble's constant: see Hubble parameter.

Hubble's law: see Hubble parameter.

inflation: hypothetical phase during the history of the universe during which cosmic expansion would have been extremely rapid (an exponential rather than a power law function of time). Such a period would have been possible if the universe had, during the course of its history, been dominated by vacuum energy*.

ionisation: a state of matter in which electrons are separated from the nuclei. The degree of ionisation measures the abundance of free electrons in relation to the number of neutral (non-ionised) atoms. Above a certain temperature, matter is always ionised. The primordial universe was totally ionised during the first million years.

kelvin or degrees absolute: unit of absolute temperature. The interval is the same as for degrees Celsius but the zero on the Kelvin scale corresponds to minus 273 degrees Celsius.

leptons: matter particles that (contrary to hadrons) do not experience strong interactions.

light travel time: when we observe a galaxy (or another astronomical object) the light received was emitted very long ago. The light travel time is the time this radiation has taken to reach us since its emission.

light-year: unit of distance, equal to the distance travelled by light in one year. It has the value 9.4605×10^{12} km.

lithium: light chemical element – the third in classification – whose nucleus contains three protons. It was largely synthesised during primordial nucleosynthesis.

Local Group (LG): this is the name given to the group of galaxies very close to our own. Amongst others, it comprises the two Magellanic clouds, the Andromeda galaxy and a few dwarf galaxies.

Local Supercluster: the huge concentration of matter that engulfs our own Galaxy, the galaxies of the Local Group and many others as well as several clusters including Virgo. It is one of the largest objects identified in the universe, and it is around 20 Mpc across.

magnetic monopole: see symmetry breaking.

manifold: this is the mathematical term given to a generalised space. Space, and spacetime, are manifolds. Surfaces are also manifolds. The simplest manifolds correspond to the space we are used to dealing with, but many more complex spaces exist. Riemannian manifolds play a very important role in general relativity and in cosmology. Among other properties, curvature plays a particularly important role. According to general relativity, space and spacetime can have curvature in the same way as a surface. The notion of curvature is fundamental in the relativistic geometry of cosmology.

matter–radiation equilibrium: during cosmic evolution, the energy density of matter dilutes less quickly than radiation. Today matter dominates. The further one goes back into the past the more radiation is important compared with matter. Matter–radiation equilibrium is defined as the instant in time when the contributions of each are the same.

metric: in order to define the properties of a manifold*, one uses a mathematical object called a metric (or a metric tensor). This notion is fundamental for calculating trajectories, distances, time intervals etc. in cosmology.

Milky Way: this is the name given to our Galaxy, and derives from the appearance of its disc seen from within. Its luminosity is due to the superposition of the luminosities of millions of stars.

muon: a particle quite similar to the electron but much heavier (and much less common). It defines one of the three families of leptons.

non-linear, linear: the word linear, in a very general way, applies to a physical problem in which the quantities are so small that one can make an approximation, called linear, that allows one to solve the equations. In cosmology this terminology applies to dyamical problems concerned with cosmic structure formation, or local condensation of matter evolving under the influence of gravity. So long as these condensations are relatively very weak, the linear theory can be used to follow their evolution. All cosmic fluctuations thus passed through a linear phase. Today most fluctuations (on the scale of galaxies or clusters of galaxies for example) have grown to such an extent that it is no longer admissible to apply this approximation; they have entered (a long time ago), the non-linear phase. Nevertheless, even today, there are fluctuations on a very large scale that probably are still in a stage of their evolution that can still be described by the linear approximation. One then speaks, in relation to them, of linear scales. When applied to spatial scales, the word linear (or non-linear) qualifies condensations greater (less) than ten or so Mpc.

parallax: parallax is the angle subtended by the Earth's orbit at a star. It provides a measure of the distance of the star, from which is defined the unit of distance called the parsec*. The method of parallaxes allows one to estimate the distances to the nearest stars.

parsec: (see parallax) a parsec is 3×10^{13} km, or 3.262 light years*.

pion: an elementary particle (also called the π meson); of mass 140 MeV.

primordial nucleosynthesis: atomic nuclei have not always existed. Nuclei of heavy elements have been produced by the stars, and those of the light nuclei were formed during a very dense and hot phase of cosmic evolution, primordial nucleosynthesis. The predictions for this phase are the main success of the big bang.

quantum chromodynamics: theory of the strong interactions. The strong interactions involve particles called hadrons* (essentially the nucleons and pions). According to quantum chromodynamics, these particles are made up of even more elementary particles called quarks* and antiquarks*, which interact via exchange particles called gluons.

quantum field theory: this is the general framework within which elementary particles and their interactions are described consistently with special relativity.

quantum statistics: all particles obey quantum statistics. Depending on whether the particles are fermions (half integer spin) or bosons (integer spin), the statistics are described by the Fermi–Dirac formula or the Bose–Einstein formula. The distribution depends on the temperature and the chemical potential, as well as on the properties of the particles (essentially the possible spin states of the particles in question).

quark–hadron transition: the primordial universe was probably filled with quarks*. During the quark–hadron transition they were brought together in the form of hadrons*, mainly protons and neutrons.

quarks: see quantum chromodynamics.

quasar: quasars are very distant astronomical objects (more distant than most galaxies). We do not yet understand how they were formed or their true nature. They are the seat of very energetic phenomena which make it possible for them to be seen at very large distances.

recombination: this is the moment, in the history of the universe, when free electrons recombine with protons to form hydrogen atoms. Matter ceases to be ionised, and decouples from electromagnetic radiation. This defines the period of decoupling, a million years or so after the big bang.

redshift: the frequencies (or the wavelengths) of galaxies are observed at values different from their values at emission. The corresponding shift is due to the expansion of the universe. This is a fundamental quantity in cosmology which describes both the distance of astronomical objects in space and the time in the past of the universe. It is related to the inverse of the scale factor*.

relativistic: matter is said to be relativistic when its thermal velocity approaches the speed of light. The temperature is so high that the kinetic energy is close to or greater than the rest mass energy.

right ascension: angular coordinate used to fix the position of the stars in the sky measured along the equator.

scale factor: all cosmic dimensions grow with cosmic time in proportion to a quantity called the scale factor which provides a distance standard.

spacetime: the objects considered by cosmology are distant both in space and in time. This requires the simultaneous consideration of intervals of time and space. Thus one uses the more general framework of spacetime. In general relativity one has to consider phenomena taking place in spacetime rather than at points in space and instants of time.

symmetry breaking: an event during which the system passes from a state having a certain symmetry to another that does not have this symmetry. According to the grand unified theories*, the universe could have undergone such an event with important consequences. According to the given variation of the model, there might have arisen very massive objects such as magnetic monopoles* or cosmic strings* with strange properties. According to other versions there might have been a period of cosmic inflation*.

tau: this is the particle representing the third leptonic family.

vacuum energy: according to quantum field theory*, the ground state of a system, also called the vacuum, can possess an energy density in the same way as a particle distribution or radiation. It is possible that the universe was, during a phase of its development, energetically dominated by such a component. This would have resulted in an era of inflation*.

Virgo: the name of a constellation. It is also the name of the nearest cluster to our own (Local Group) and is seen in this constellation.

weak interactions: one of the four types of interaction between particles. Today they are understood to be similar to electromagnetic interactions since they can both be described in the framework of the 'electroweak theory'. Nevertheless, they do not involve the same particles. Neutrinos experience weak interactions.